T0201874

Genetic Reconstruction of the Past

Genetic Reconstruction of the Past

DNA Analysis in Forensics and Human Evolution

HENRY A. ERLICH

OXFORD
UNIVERSITY PRESS

Oxford University Press is a department of the University of Oxford. It furthers
the University's objective of excellence in research, scholarship, and education
by publishing worldwide. Oxford is a registered trade mark of Oxford University
Press in the UK and certain other countries.

Published in the United States of America by Oxford University Press
198 Madison Avenue, New York, NY 10016, United States of America.

CIP data is on file at the Library of Congress

ISBN 978–0–19–767536–6

DOI: 10.1093/oso/9780197675366.001.0001

Printed by Integrated Books International, United States of America

"The past is never dead. It's not even past."

William Faulkner

For Justin, Lucan, Niobe, Thaisa and their progeny . . . present and future.

Contents

Preface

Thanks to the development and application of DNA technology, the way we think about human history and the way we now consider criminal evidence and investigate crimes, ranging from individual murders and rapes to mass atrocities, has been dramatically transformed. In the last thirty-five years or so, DNA analysis has revolutionized the seemingly disparate fields of evolutionary biology and forensic science, providing precise and quantitative answers to old questions and raising entirely new ones. What those fields now share is the recently developed capacity to use DNA sequence information to make inferences about both distant and recent historical events. DNA analysis of forensic biological specimens can help reconstruct the events that transpired at the scene of a crime, and similarly, analysis of samples from different human populations or from other species, alive or extinct, can now help reconstruct the ancient past, the historical relationships among contemporary and extinct human populations. In this book, I tell the story of how the DNA technology my colleagues and I developed in the mid-1980s was applied to questions of guilt and innocence in the criminal justice system as well as to the mysteries of human evolution.

I had always been deeply interested in both of these areas, but my opportunity to actively work in them didn't begin until 1986, starting with collaborations with Ed Blake, a forensic scientist who worked in the same building I did in Emeryville, California, and with Ulf Gyllensten, an evolutionary biologist from Sweden, who joined my lab to do a postdoctoral fellowship. I had not been trained as either a forensic scientist or an evolutionary biologist but as a molecular geneticist, later developing and applying DNA technology to these two fields. The technological breakthrough that has made DNA analysis so powerful and informative in these fields was the development by my colleagues and me at Cetus Corporation, a small biotech company in the San Francisco Bay area, of the polymerase chain reaction (PCR) method for amplifying specific DNA sequences. PCR is capable of synthesizing millions of copies of a specific targeted DNA fragment in a test tube using a process similar to the way DNA replicates in the cell; the sequence of the DNA "amplified" from a tiny blood spot or a bone fragment could then be analyzed and used to make inferences in both forensic cases and evolutionary studies. PCR used to be a relatively arcane acronym, familiar only to biological researchers and clinicians. Now, in the pandemic era of nasal swabs and covid testing, we are all familiar with the term, if not the technology.

Part I discusses the development and introduction of the PCR method of DNA amplification and its application to some forensic specimens in collaboration with Ed Blake of Forensic Science Associates and the subsequent development, spread, and establishment of these techniques in the US and international judicial systems. I discuss controversial historical and current issues in forensics DNA analysis in the context of specific criminal cases and my personal involvement in some of these groundbreaking cases. **Part II** discusses the application of PCR to analysis of the genetic relatedness of different human populations as well as of archaic species (e.g., Neanderthal) to illuminate their historical relationships.

The genetic reconstruction of the past, the use of DNA sequence information to make inferences about historical events, is a metaphorical scaffold on which I hang my reflections, discussions of individual cases and controversies as well as the occasional scientific explanation. In the pages that follow, I have tried not only to show the remarkable findings that have transformed the criminal justice system and our newly enhanced understanding of our human heritage but also to impart the flavor of "doing science," identifying some features that distinguish scientific research from other human activities. In these discussions, I try to make explicit the kind of assumptions used in the interpretation of DNA matches in forensics, in the calculation of probabilities, and in the construction of evolutionary trees. Like most accounts in which the author has played an active role, this book has its share of back-patting, name-naming, and the occasional riding of hobby-horses. Although trained as an immunologist, I am not totally immune to these aspects of the genre, but I have tried to focus on issues of general interest in the discussion of DNA analysis in forensics and evolution and to present the issues as fairly and clearly as I can.

A common phrase indicating genetic inheritance is "he/she came by it *honestly.*" (Does this usage imply that learned traits are somehow *dishonest*?). Since my last name means "honest," perhaps I was predisposed to genetics research. For people with a genetic predisposition for some outcome—say, an autoimmune disorder like type 1 diabetes, a disease I have studied for many years—an environmental trigger, like a virus, is often required to elicit autoimmunity and disease. In my case, the trigger that sparked an interest in genetics was an inspiring biology course taught by a charismatic teacher. In my freshman year at Harvard, I had chosen to major in Renaissance history and literature when, to satisfy a distribution requirement, I enrolled in a general biology course created and lovingly taught by George Wald. As part of his engaging and often thrilling introduction to biochemistry, genetics, and developmental and evolutionary biology, Wald would quote James Joyce and Gertrude Stein along with Louis Pasteur and Charles Darwin. He made molecular biology seem an essential and fundamental part of philosophy and much more exciting than the actual/nominal philosophy classes I was taking at the time. He later received the Nobel Prize (1967) for his pioneering work on the biochemistry of vision, but to his enthralled students, he was simply our inspirational teacher.

Later in my career, I had an opportunity to hear Wald's lecture again and found him still inspiring, if a bit melodramatic. To concerns that the approach of molecular biology to the study of life was overly reductionist, he responded, "I say this not to denigrate man but to exalt the molecule." A bit much, perhaps, for a graduate student, but pretty seductive and heady stuff for a freshman and ultimately a philosophical perspective I still accept. The successful reductionism of molecular biology, as well as of other scientific enterprises, need not diminish the wonder and awe with which we experience the natural world.

The giant polytene chromosomes of *Drosophila* (the common fruitfly) I gazed at through a microscope in the lab for Wald's course were beautiful and intriguing, but it was really the logical elegance of the recently published papers on the genetic code and on gene regulation in the *lac* operon of *E. coli* by Francois Jacob and Jacques Monod that sealed the deal. These two heroes of the French Resistance were also my first two scientific heroes. They demonstrated that fundamental insights into how genes were turned on and off in response to environmental stimuli could be achieved simply by observing the behavior of a few mutant strains of *E. coli* and by constructing a theoretical model of interacting proteins and genes that governed gene expression. This model proposed a mechanism by which the exposure of bacteria to a particular sugar "induced" the synthesis of enzymes capable of metabolizing that sugar to support bacterial growth. The power of abstract logic and reason and the beauty of genetic analysis in generating these insights were both exhilarating and enticing. In fact, the entire field of genetics was created by careful and precise observation coupled with a logical model in Gregor Mendel's transformative studies of peas in his abbey garden in Brno, Slovakia. The fundamental truths of inheritance were first revealed by the good abbot's simply counting round and wrinkled peas and interpreting the results.

I switched my major to biochemistry and signed up for an upper level genetics course. The course was taught by the legendary geneticist Matt Meselson, one half of the famous Meselson-Stahl experiment that revealed how DNA replicates and that, as an illustration of beautiful experimental design,[1] is still an essential part of every introductory biology curriculum.

Later, as a graduate student in the Genetics Department of the University of Washington in the late 1960s, I was fortunate to work with Jonathan Gallant, a young bacterial geneticist, as my PhD advisor. Jon was an eccentric but brilliant scientist with a wicked sense of humor and a decidedly counterculture orientation; he had a radio show on a local FM station and wrote incisive, provocative pieces in local alternative media. Jon made it seem that one could be a good, serious scientist and still be engaged in the then very exciting and turbulent world outside the lab. As a student fascinated with the problem of gene expression but also an active member of Students for a Democratic Society and an occasional writer for the Seattle underground newspaper, *The Helix*, I decided that Jon's lab, which studied the regulation of genes in *E. coli*, was the right choice for me.

Before committing to Jon's lab, however, I decided to take a break from grad school and joined VISTA (Volunteers in Service to America) and chose to work with a program with street gangs in Santa Fe, New Mexico. It was, after all, the 60s, and this was my first significant departure from the student experience. My choice was fortunate in many respects. I loved the natural beauty of northern New Mexico and the vibrant Spanish and Native American cultures so evident in Santa Fe; I made some lifelong friends, notably Godfrey Reggio, a philosophically minded Christian Brother teacher and community organizer, the charismatic founder and leader of a program for local street gangs, Young Citizens for Action, and later, an avant-garde filmmaker (*Koyaanisquatsi*). My year in New Mexico resulted in a much broader perspective and, when I returned to grad school, a renewed commitment to basic scientific research but also to seeing genetics research in a wider social context.

---

[1] The word scientists use to describe a particularly well-designed experiment that is simple yet definitive is "elegant."

Following graduate school, I arranged a postdoctoral fellowship with Ted Cox in Princeton, studying a specific gene that influenced the mutation rate in bacteria. Ted had recently published an elegant paper showing how natural selection can operate on the mutation rate, that is, the rate at which new genetic variants were produced. Although I had always been interested in evolution and natural selection, Cox's work illustrated the value of an experimental and quantitative approach to evolutionary studies. While at Princeton, I also took a course in immunology, a subject I found fascinating and, at the time, still fairly mysterious. This course ultimately changed the direction of my professional life, but it was a conference at Franconia College, in New Hampshire, that most profoundly transformed my life. I met my future wife, Brenda Way, there.

Intrigued by the mystery as well as the potential clinical applications of immunology, I decided to take a second postdoc fellowship in immunogenetics and had the good fortune to be accepted in Hugh McDevitt's lab at Stanford in 1975, another move that ultimately contributed to my DNA forensic work. The genes that controlled the immune response had just been identified by McDevitt and colleagues using inbred strains of mice that either responded or failed to respond to a specific antigenic stimulus. The mechanism underlying this genetic control of immune responsiveness was, at the time, one of the most fundamental and intensively researched problems in immunology. My initial project was to clone some of the genes in the genomic region known as the major histocompatibility complex (MHC) using the newly developed technology of recombinant DNA developed by Stan Cohen, whose lab was just down the hall, and Herb Boyer at University of California, San Francisco.

In mice and man, as in virtually all animals, the genes in the MHC are very polymorphic, that is, highly variable among the individuals in a population. In humans, the MHC is known as the HLA (human leukocyte antigen) region, and some of the genes in this region have, we now know, thousands of genetic variants (alleles). It was one of these highly polymorphic HLA genes that, 10 years later, my lab at Cetus used to carry out the first forensic application of DNA analysis, as described in Chapter 1. In the immune system, these genes function to distinguish self from nonself and are the genes that require donor–recipient matching in bone marrow and solid organ transplantation. Although my research at Stanford focused on mice, it was clear that analyzing this genetic region in humans would be enormously interesting and potentially clinically—and, as it would prove, forensically—valuable.

It wasn't until 1979, though, when I moved to Cetus Corp that I began to focus on the HLA region and to try to analyze the polymorphism and genetic organization at the DNA level. At the time, the HLA typing used to match donor and recipient for transplants was based on reactivity with antibodies.[2] When I came to Cetus, I had the naively ambitious goal of developing an HLA DNA typing system, although many people in the field tried to dissuade me, pointing out that serology, the HLA typing system based on antibodies, seemed to work just fine. Now, however, virtually all HLA typing is based on PCR amplification and DNA analysis. This HLA project seemed to satisfy the commercial requirement at Cetus to work on something with potential clinical utility (e.g., transplantation, disease association, paternity and forensics testing) as well as to address some fundamental genetic research issues.

I had accepted the position at Cetus, in large part, because of Brenda. She had left a tenured position at Oberlin College to come west to live with me in San Francisco. She arrived with her three kids and twelve dancers, members of the Oberlin Dance Collective she had founded a few years earlier, in a big yellow school bus. So, when my postdoc at Stanford was over, my priority was to find a job in the Bay Area. I had presumed I would end up in an academic institution, doing basic research. My father and uncle were both professors (literature and economics), and university life was what I knew and liked; it was, in a sense, a default option. But academic openings in the Bay area were slim, and I had been hearing good things about these new institutions, small start-up commercial companies doing interesting and innovative biological research. Cetus invited me to give a lecture and offered me a position. I anguished over whether I would be going over to the Dark Side by working for a for-profit company, but I was reassured by the fact that two of my Stanford mentors, Hugh McDevitt and Stan Cohen, were advisors and, most importantly, by spending time with many of the bright, young Cetus scientists, notably Tom White and David Gelfand, who had recruited me. Their energy and enthusiasm were palpable, and the culture seemed decidedly more "counter" than "corporate."

And so, I found myself in 1979 surrounded by other young and eager scientists, hoping we could find something interesting and useful to work on,

[2] The antibodies used for HLA typing had been generated in women who, following multiple pregnancies, had been immunized against the HLA molecules of the father, or in blood transfusion recipients who had been immunized against the HLA molecules of the blood donor. This system was developed by Jean Dausset (France), Jon Van Rood (Netherlands), and Rose Payne (United States). Dausset, along with two other pioneers of immunogenetics research, George Snell and Baruj Benacerraf, received the Nobel Prize in Medicine in 1980.

something intellectually and economically sustaining that would help Cetus survive and, with luck, thrive. I switched my research focus from mouse genetics to human genetics and, as it turned out, my interest in developing DNA technology to analyze clinically relevant genetic variation proved to have important consequences for both forensic applications and evolutionary studies.

One of my responsibilities and preoccupations at Cetus was to try to imagine potentially useful and therefore, presumably, commercially valuable, applications of the basic genetic research ongoing in our lab. This research involved a search to identify genetic variation with clinical relevance and develop techniques to "genotype" samples—that is, identify the genetic variants carried by an individual, with the hope of finding specific variants that might, for example, confer increased or decreased disease risk to the individuals carrying them. The analysis of genetic variation in human populations, however, turned out to have two other significant applications; these are the subject of this book.

Acknowledgments

Before the book was the work, the initial PCR team at Cetus trying to amplify the beta-globin gene was Fred Faloona working with Kary Mullis, and Randy Saiki and Steve Scharf working with me and Norm Arnheim. David Gelfand, Suzanne Stoffel, John Sninsky, Shirley Kwok, and Russ Higuchi all made critical contributions. As Vice President of Research at Cetus, Tom White provided invaluable support in the face of skepticism from colleagues and outright hostility from our President. The forensic applications of PCR (Part I) would not have happened without Ed Blake and Ulf Gyllensten, and Steve Mack helped carry out the evolutionary analyses (Part II).

During the long gestation of this book, Jonathan Cobb provided experienced and helpful editorial feedback and guidance, constantly encouraging me to make it accessible to a general reader. ("Explain or delete" was a recurring theme.) Any failure to achieve this goal is, as the late lamented Jimmy Buffet sang, "my own damn fault."

I'm very grateful to Tom White, who was unique among reviewers in that he knew all the characters, the history, as well as the science; Tom's feedback on the whole book was invaluable. Eric Stover's work on human rights has been an inspiration and his review of Chapter 8 was extremely helpful. Arielle Bransten, a very bright high-schooler, helped assemble the bibliography and provided feedback from the perspective of a general non-technical reader.

Many thanks to my agent, John Thornton, and to the immensely helpful team at Oxford University Press, Jeremy Lewis, David Lipp, and Mahalakshmi Bulamurugan. Finally, for her support and insightful editorial feedback, I thank my wife, Brenda Way, my partner in all things that matter.

Introduction

The Stories DNA Tells

- The DNA profile of biological evidence found at the scene of 5 different murders in Southern California matched that of John Thomas, Jr, a 72-year-old Los Angeles insurance adjustor and convicted rapist. Police officials said DNA evidence might link Thomas to as many as 25 other unsolved regional murders and sexual assaults in the 1970s and 1980s.
- DNA analysis of human remains found in a southern Utah desert identified the bones as those of Everett Ruess, a young and promising poet and painter who disappeared in 1934.
- A University of Pennsylvania research team, led by Dr. Sarah Tishkoff, analyzed the DNA of sample populations from Africa and the rest of the world and concluded that modern humans originated from south-western Africa and migrated out of Africa some 50,000 years ago from the African coast of the Red Sea.

The common thread that links these disparate stories—and the topics in this book—is the use of DNA testing and population genetics data. Coincidentally, these stories, which illustrate this book's organizing principle, all appeared in the same issue of the *New York Times*, on May 1, 2009. Now, 15 years after that May Day issue and over 35 years since the first criminal case in which DNA analysis was entered as evidence (*Pennsylvania vs. Pestinikas*, 1986) and the first exoneration based on DNA evidence from a person wrongly convicted (*Illinois vs. Dotson*, 1989), the historical secrets embedded in the DNA sequence are being revealed by scientists around the world—from the analysis of bone fragments in mass graves to the fossilized remains of archaic species.

As Jill Lepore points out in her magisterial history of the United States, *These Truths* (2018) "history is the study of what remains, what is left behind." For the reconstruction of crime scenes or human migrations, it is DNA that "is left behind" and tells the story. In the last few years, serial killers have been identified by searching DNA databases; in one recent case (the Golden State Killer, 2018), the database that led to the suspect was designed not for purposes of law enforcement but for tracing genealogy and ancestry. Meanwhile, new advances in DNA-sequencing technology, applied to the analysis of both living and dead humans, now make it possible to track the human migrations that created the biogeographic composition of the world we know today.

Three fundamental properties of DNA make analysis of this molecule so valuable in forensic science and evolutionary studies. As befits the molecule that encodes biological inheritance, DNA is chemically stable. Unlike other macromolecules like proteins, RNA, or lipids, DNA can survive essentially unchanged for thousands of years. Some tissues protect DNA more than others (teeth are good in this regard; soft tissue, not so much). The DNA from ancient specimens or from forensics samples subjected to environmental influences can be "degraded"—present only as short DNA fragments—and suffer some minor chemical changes, but the essential information, the sequence of chemical bases—adenine (A), cytosine (C), thymine (T), and guanine (G)—often remains intact and can be recovered or "reconstructed" from analysis of the samples.[1]

The second feature of DNA critical for all of these studies is the variation that occurs in the sequence of bases among species and among different individuals of the same species. The human genome has just over three billion bases. Over time, mutations (e.g., base substitutions, as in an A to G change) accumulate in DNA, resulting in a pattern of base changes such that the DNA sequence of a given gene in different species will vary in a systematic way that allows the reconstruction of their evolutionary history. In fact, the *extent* of the sequence differences between these species can be used to estimate the time of their divergence from a common ancestor. Among humans, well over 99% of the genome sequence is identical, but focusing on the less than 1% that is variable among individuals—that is, "polymorphic"—provides a measure of genetic individuality (with the exception of "identical" twins)[2] that has proved enormously powerful in forensic application of DNA analysis.

---

[1] The analysis of ancient DNA has revealed that a small proportion of the cytosines have been converted into thymine. This property can be used to distinguish ancient DNA from modern DNA.

[2] Recently, sequencing the whole genome of identical twins has revealed some DNA differences even in these cases as well (*Commonwealth of Massachusetts vs. Dwayne McNair*, 2014).

The third feature is that DNA, unlike protein, can be "amplified," allowing the sequence of even a single DNA molecule to be experimentally determined. If a single protein molecule were present in a sample, no analysis would be possible; the amount would be grossly insufficient for performing the necessary tests. However, if only a single DNA molecule were present, millions of copies of a particular DNA sequence could be synthesized in a test tube using a method my colleagues and I reported in 1985 (the polymerase chain reaction, or PCR), and, following such copying (amplification), analyzed and interpreted. It is this capacity, the ability to generate a million copies of a "needle" in the classic "needle in the haystack" problem, that has transformed both forensic science and evolutionary biology. (Before the COVID-19 pandemic, PCR, which we initially called "enzymatic amplification," was a familiar acronym to biologists and the medical community but not to the general public. Now, after all of us have had our noses swabbed repeatedly, we are all only too familiar with PCR.[3])

The connection between DNA analysis in forensics and in evolution is, of course, much more fundamental than simply my personal interest and my own involvement in both of these areas. The inferences in these two fields are made from basically the same kind of genetic data, generated by the same DNA technology, the amplification and analysis of specific DNA genomic segments by PCR. In both fields, hypotheses are tested by comparing DNA sequences in regions of the genome whose sequence varies among individuals. In both fields, it is the frequency of these sequence variants (alleles) in various populations that allows us to evaluate the significance of these comparisons and to test the hypotheses specific to each area: Is blood stain A derived from the suspect or is population A genetically related more closely to population B than to population C?

The first forensic application of DNA analysis in the United States was in *Pennsylvania vs. Pestinikas*, a murder case; my lab carried out the analysis in collaboration with Ed Blake of Forensic Science Associates (FSA), as mentioned in the Preface. Since our initial work at Cetus applying PCR to the analysis of a few autopsy samples from the Pestinikas case that summer of 1986, millions of samples have been analyzed and many thousands of cases have been resolved using what are basically the same DNA techniques. The identification of human remains, from the 9/11 World Trade Center attack to

---

[3] I realized that PCR had moved beyond the community of molecular biologists when I read in the opening pages of Richard Power's novel *Orfeo,* that Power's protagonist, a composer, was carrying out PCR in his garage lab.

mass graves in the Balkans and in other humanitarian disasters, relied on this same technology. The subsequent development, spread, and establishment of these techniques have transformed criminal justice systems in the United States and around the world. Within the United States, DNA testing has raised challenging and difficult questions about the death penalty as well as about the reliability of other forms of evidence, such as eyewitness testimony and confessions and forensic techniques such as hair morphology and tooth bite analysis. The prospect of identifying the perpetrator in crimes without a suspect (cold cases) by checking an evidentiary genetic profile against a database of genetic profiles of convicted felons (or, in some cases, arrestees) has already become standard practice, although exactly how this should be done remains controversial. Now, just over three decades since the Pestinikas case, it is hard to imagine our justice system without access to the DNA testing of biological evidence.

The same DNA technology that allows the analysis of a shed hair (i.e., one with no root) or tiny blood spot at a crime scene also enables sequencing the DNA of a bone fragment from a Neanderthal skeleton or another archaic species. But the connections between forensics applications and evolutionary studies go far beyond shared DNA technology and genetic population data. These two fields share a common goal: solving historical mysteries. In a sense, all scholarly research is detective work but, as the philosopher Karl Popper pointed out, science differs from many other intellectual pursuits in that, to the extent possible, it requires testable (i.e., falsifiable) hypotheses.

Unlike many sciences, forensics and evolutionary biology are essentially historical disciplines, without access to direct experimentation to test hypotheses about the past. However, experiments designed to test such evolutionary mechanisms as mutation and natural selection have proved valuable and illuminating. In the past few years, the ability to recover and analyze the genomic sequences of ancient specimens, notably bone remains of archaic species like the Neanderthals, has provided previously inaccessible evidence for constructing and testing historical hypotheses. And even when direct experimentation is not available, competing models and hypotheses strive to account most simply and completely for a given set of historical data—a struggle for survival of the "fittest" hypothesis—that is, the one that best fits the data.

One critical aspect of falsifiability, built into the design of scientific experiments that evaluate a given hypothesis, is the "control." The control component in an experiment is, in essence, a falsifiability check, a kind of institutionalized skepticism designed to provide reassurance that the experiment is working as designed and that the tentative conclusions attributing a particular outcome to a specific cause are valid. A "positive" control often involves running the experiment on a sample that should give an expected and predetermined outcome. If it does not, this result suggests that something in the experimental system is not working as assumed and whatever results are obtained may not be reliable. Another kind of control designed to experimentally test a specific hypothesis is the "control group." In a vaccine trial, a timely example as I write in the throes of the COVID-19 pandemic, the immune response of a vaccinated group is compared with the response of a "control" group, either unvaccinated or inoculated with some other vaccine. The hypothesis to be tested is that the COVID-19 vaccine will elicit a stronger immune response and confer stronger protection against some quantifiable metric (e.g., levels of specific antibody, infection, hospitalization, death) than the "control" vaccination. Here, the alternative hypothesis is the "null" hypothesis, namely that there will be no difference in immune response between the two groups.

Yet another kind of control tests a specific alternative hypothesis to account for the data. My favorite example of testing alternative hypotheses is a brief conversation between the two main characters in *Rosencrantz and Guildenstern Are Dead*, the existential comedy Tom Stoppard created based on *Hamlet*. The two characters are deeply puzzled (*Why are we here? Who called us?*) and fundamentally confused about their own identities—they don't even know who is who. One character attempts to resolve this confusion by addressing the other as "Rosencrantz?" and the other responds. The puzzle is thus apparently resolved with the speaker then identified as Guildenstern and the responder as Rosencrantz. The speaker then proceeds to run a "control experiment" and addresses the other, once again, as "Guildenstern?" and the other responds affirmatively again. Initial hypothesis rejected: back to square one, and the existential confusion of identity. This kind of control, then, is simply a way of trying to reject/falsify one's hypothesis. The attempt to consider alternative explanations, institutionalized in the concept of the "control," is one of the distinguishing elements of the scientific enterprise. The design of experiments and the analysis of results are meant to consider the possibility of mistaken interpretations although most scientists are not quite as melodramatic as Oliver Cromwell in his famous pleas in 1650 to the Church of Scotland, "I beseech you in the bowels of Christ, to think it possible that you may be mistaken."

Human nature is such that scientists typically hope that the results from a given experiment will be consistent with their favorite hypothesis. Acknowledging this psychological truth, a standard self-deprecating joke, in response to the results of a control experiment that rejects the favored hypothesis, is that the experimenter "ran one too many controls." The late Sydney Brenner, the charismatic and elfin geneticist who helped "crack" the genetic code, coined the term "Occam's Broom,"[4] to describe the desire "to sweep under the carpet what you must to leave your hypotheses consistent."

In a forensics context, the relevant hypotheses are not the general scientific ones Popper envisioned but particular ones related only to a given case. The hypothesis that a given DNA sample came from the suspect can clearly be rejected (falsified) if the genetic profiles of sample and suspect are different. This result is known as an "exclusion," meaning the suspect is *excluded* from the universe of individuals who might have left that evidence sample. Like a fingerprint from the crime scene that doesn't match the suspect, a different genetic profile means, without any need to resort to statistics, that the evidence did not come from the suspect. However, if the genetic profiles "match," this outcome requires careful and precise interpretation in both words and numbers. When the profiles of the evidence and reference samples are indistinguishable, this result is often referred to as an "inclusion" or a "failure to exclude"—or, to be even more conservative, as forensic analysts are encouraged to be on the witness stand, one might testify that the evidence profile "is consistent with" that of the suspect.

What really matters is not the language used but the numerical significance of the inclusionary result, typically expressed as the random match probability (RMP)—that is, the probability that the DNA profile of some individual chosen at random in the relevant population would also "match" the evidence. The probability of choosing at random any particular genetic profile is related to the frequency of that profile in the relevant population. An RMP of 1/100, *in the absence of any other evidence,* may not be very compelling evidence, while an RMP of 1/100,000,000 argues very strongly that the evidence sample came from the reference individual.

The significance of a match also depends on the universe of potential perpetrators. When testifying in a trial of a suspect charged with the murder of a New Jersey State Trooper, I was asked by the judge, if the RMP was 1/

[4] Brenner's joke refers to "Occam's Razor," the logical principle, attributed to the medieval philosopher William of Occam (or Ockham) that, when considering competing hypotheses, advocates choosing the simplest one, the one with the fewest assumptions.

1 million, didn't that mean that there were around 300 people in the United States (population of around 300 million) who would also match the profile? I responded that his calculation was correct, but that most of them would not have been on the NJ Turnpike at the time of the trooper's shooting. The RMP metric raises the issue of what *is* the relevant population to consider and, of course, requires a population database for estimating the frequency of any particular genetic profile. It is the ability to generate numerical estimates of RMP based on scientific evidence and statistical calculations that distinguishes DNA inclusions from "matches" obtained with other kinds of forensic evidence such as fingerprints or hair and bite-mark morphology.[5] These forms of pattern matching are not based on rigorous population frequency data and, therefore, claimed inclusions are not associated with meaningful probability or statistical estimates like the RMP or another common way that the DNA evidence of an inclusion is presented to the jury, the likelihood ratio (LR), a metric that compares the probability of the evidence given two competing hypotheses.

In the forensic context, the prosecution "hypothesis" in the case of an "inclusion" is that the DNA evidence came from the suspect and, furthermore, that this result implicates the suspect in the commission of the crime. However, the defense "hypothesis"—that is, "Some Other Dude Did It"—will, typically, still be maintained because, in our adversarial judicial system, that is the job of the defense, and because alternative explanations for the matching DNA profiles, like coincidence—the source of the evidence sample is someone who happens to have the same profile as the suspect, resulting in an adventitious match—can't always be ruled out. Other explanations of inclusion might be lab error or contamination. An alternative response by defense attorneys might be to concede the match and the source of the evidence but to invoke evidence planting, as we'll see in Chapter 7 on the OJ Simpson trial, or in the case of a rape charge, consensual sex, as we'll see in Chapter 2.

The application of DNA analysis is more pragmatic and narrowly focused and less classically "Popperian" in the sense of formal hypothesis testing in forensics than it is in evolutionary studies or in other biological research. DNA analysis in evolutionary research might address questions of how closely related are modern humans and Neanderthals and did they

---

[5] The reliability of hair morphology and bite-mark analyses has been called into question by the recent report of the Presidential Council of Advisors on Science and Technology (the PCAST Report, 2016).

interbreed. DNA research in forensics, however, tends to be more involved in technology development and validation and in issues of how to interpret matches/inclusions than is research focused on the more fundamental questions asked in evolutionary studies. The greatest overlap between the applied discipline of forensics and the more basic research field of population genetics is in the area of statistical evaluation of inclusions (matches). How best to estimate RMPs has been and continues to be among the most scientifically interesting and judicially consequential questions of forensics DNA analysis.

A vital aspect of any forensic investigation or evolutionary study is the practice, where feasible, of *replication* of results obtained. When it comes to experimental work, the reason that scientific papers (ideally) include a detailed description of the experimental methods used is so others can repeat the experiments. If the published results cannot be replicated by others, then the conclusion inferred from those initial results will, with some caveats, be rejected or, at least, questioned. Such efforts at replication constitute an essential self-correcting aspect of the scientific enterprise.

Forensics and evolutionary genetics, as noted above, are both science-based yet essentially historical studies in that they attempt to explain the past. The desire to solve historical mysteries, of course, is much deeper and more basic than the 17th-century emergence of the scientific method. The desire to construct narratives, whether stories, histories, myths, or legends, about human origins seems universal. Virtually every human culture has elaborate creation myths, and contemporary societies still retain a fascination with origin narratives, even if they are now more often based on genetics as well as data from archeology and paleoanthropology. Commercial firms like Family Tree, Ancestry, and 23 and Me have grown into large successful businesses by offering ancestry accounts based on DNA analysis of cheek swabs or saliva.

Another fundamental feature of human communities is the value placed on justice, stemming from what seems like an innate sense of fairness and morality. The classic experiments with capuchin monkeys and chimpanzees by the well-known primatologist Frans DeWaal speak to a concept of "evolved morality."[6] If we accept DeWaal's notion of "evolved" morality, a sense of fairness may be embedded in our genomes. The analysis of the

---

[6] As a reward for completing a task, two monkeys were provided either grapes or cucumbers. Their response has been interpreted as a demonstration of an innate concept of fairness.

sequence variation in DNA to address these issues of justice as well as to illuminate our history is the focus of this book.

In recent years, the concept of DNA itself seems to have acquired a deep cultural resonance, a celebrity status not accorded to other less glamorous macromolecules like proteins, reflected in the all-too-common marketing phrase "it's in our DNA."[7] In addition, this comforting notion of individual uniqueness combined with the potential connection to the past and to extended family members may account, in part, for present-day social fascination with its current uses in the realms of history and criminal justice.

The desire to solve mysteries unites these two disciplines but it is, ultimately, the passionate wanting to know, what Einstein referred to as "a holy curiosity," that underlies the scientific enterprise and, indeed, all scholarship. My favorite expression of this very human passion is the one Tom Stoppard, in his novel *The Invention of Love*, puts in the mouth of A. E. Housman, the poet and scholar: "The real thing, which is to shine some light, it doesn't matter on what, it's the light itself, against the darkness, it's what's left of God's purpose when you take away God."

Some writer whose name I can't remember made a distinction between *mysteries* and *puzzles*. Puzzles are the realm of questions that have a simple answer that we strive to identify while mysteries are more complex areas where the answer is neither simple nor known and we struggle incrementally toward a better understanding. If we accept this distinction, then in forensics, we try to solve a historical puzzle and identify the perpetrator of a given crime. In trying to understand the complex evolution of human populations and human origins, however, we are dealing with a deep mystery. (In this lexicon, readers of police procedurals and detective novels may thus be surprised to learn that they have been immersed not in *mysteries* but in *puzzles*.) This distinction notwithstanding, the urge to ask and resolve these historical questions seems an intrinsic part of human nature, as does the desire to solve crimes and seek justice. The recent emergence of DNA analysis has transformed our ability to address both of these universal human needs, as we'll see in the pages that follow.

---

[7] The preoccupation with genomic sequences has recently entered the intersection of artistic production and blockchain technology. In her exhibition at the Berlinische Galerie, Berlin, the Polish-German artist, Alicja Kwade, has printed out her entire genome sequence on 259,025 sheets of paper sheet hung from wall to ceiling, highlighting in bold text, the sequence variants (less than 0.1%) unique to her. This work, titled *Selbstportrait*, has now been transformed into a series of 10,361 NFTs, each consisting of a 25-page pdf.

PART I
RECONSTRUCTING THE CRIME SCENE

1

All Rise

DNA Enters the Courtroom

The headline for a *New York Times* article some years ago (September 27, 2006) asserted that the "CSI" DNA techniques familiar to many TV viewers the world over were just now beginning to be applied to investigations of elder abuse and neglect in nursing homes. As it happened, a nursing home investigation is where forensic use of this technology started over 35 years ago, in 1986. *Pennsylvania vs. Pestinikas*, in which the first use was made of DNA analysis in a criminal case, involved the prosecution of a nursing home for the allegedly negligent death of a patient.

Pennsylvania vs. Pestinikas was, in many respects, an unusual and unlikely case for the introduction of DNA evidence using a novel technology into the US judicial system. The case involved a suspected homicide in which the victim, Joseph Kly, a 92-year-old retired coal miner, was found dead in the fall of 1984 in the Stagecoach Inn, a rest home and rooming house where he had been living. The inn was operated by Walter and Helen Pestinikas, respected members of the Scranton community, who had agreed to look out for Kly's welfare and, when the time came, to arrange his proper burial. Mr. Pestinikas was also a funeral home director and responsible for the burial of Kly's remains. The prosecution sought to prove that the couple had neglected the victim's feeding and medical care and had drained his bank account as well. The results of the autopsy examination by Robert Sinnenberg, MD, conducted shortly after Kly's death indicated that he was extremely malnourished and dehydrated; the body weighed only 65 pounds.

In the fall of 1985, Walter and Helen Pestinikas were arrested on a homicide charge, and the Scranton prosecutor had the body exhumed and reautopsied. The second autopsy, conducted by Isadore Mihalakis, MD, found results for many of the tissue specimens that seemed incompatible with those of the first autopsy. In particular, the nature of the observations about the intestine and kidney specimens suggested a potential switch. Based on these observed inconsistencies, the prosecution developed a hypothesis that some of the tissue specimens had been replaced or tampered with between the first and second autopsies to hide evidence of elder abuse. To test this hypothesis, formaldehyde-preserved tissue samples from both autopsies were sent by the prosecutor to a forensics lab to determine whether the specimens from the first and second autopsy came from the same or different persons. The forensics lab happened to be just down the hall from my lab.

In the mid-1980s, Ed Blake, a forensics scientist at the aptly named Forensics Sciences Associates (FSA), rented a small office and lab in the old brick building that housed Cetus, the California biotech company where I worked from 1979 until its acquisition by Chiron, in 1991. (As is so often the case in the food chain of biotech, Chiron was subsequently swallowed/acquired by a yet larger company, the giant Swiss pharmaceutical company Novartis.) FSA consisted of Ed, a feisty and fiercely independent scientist in his 30s with a doctoral degree in forensic science from UC Berkeley, a secretary, and one technician. Ed did not suffer fools gladly, and his independence and bluntness may have prevented him from getting all the credit due to him as a forensics DNA pioneer. We used to run into each other in the hallway on a regular basis (probably not the first time a full bladder led to a productive scientific interaction); we would chat about our various projects (mine, research; his, casework) and, more often than not, discuss the potential application of the DNA methods my lab was working on to genetics-related forensics issues, such as determining paternity and identifying the source of criminal evidence samples.

In the summer of 1986, during one of these hallway meetings, Ed presented me with a serious technical problem he was confronting in the Pennsylvania case. The formaldehyde-preserved tissue samples (kidney, lung, liver) from the two different autopsies that Ed had been sent by the prosecutor could not be analyzed by conventional methods of protein analysis because the proteins in the specimens had been denatured by the formaldehyde treatment to which they had been subjected.

In 1986, when Ed approached me about the *Pestinikas* case, protein gel electrophoresis, a method of distinguishing polymorphic proteins based on their physical properties such as size and/or electrical charge and serology, a method based on the recognition of specific protein polymorphic variants by antibodies, were the only available and accepted techniques for analyzing biological evidence samples along with blood group typing (e.g., the ABO system) The fundamental idea underlying the use of genetic typing to identify the source of a biological specimen (blood, semen, hair, skin, bone, saliva) at a crime scene is to use a genetic "marker," whether protein or DNA, that varies among individuals and ask whether the pattern for this marker obtained by analyzing the specimen was the same as—"matched"—the pattern derived from a sample from a reference individual, such as the suspect or victim. The significance of such a "match" can, however, vary widely. Based on data indicating the frequency of this pattern (genetic profile, when DNA markers are used) in the population, one can estimate the statistical chances that a randomly selected individual would also match the reference sample and thus interpret the weight of the evidence, that is, the statistical significance of the observed match.

When the genetic profile of a potentially informative specimen (e.g., semen stain) is indistinguishable from that of a reference sample (e.g., blood from a suspect) based on a particular test, the result is often referred to in forensics circles as an "inclusion," as mentioned in the Introduction, a presumably less loaded term than "match." This usage raises the relevant statistical issues. What *is* the frequency of this particular genetic profile in a particular population? How is the frequency estimated? Since the frequency of this genetic profile is likely to be different in different ethnic groups or geographic regions, in what populations should the frequency be estimated? An *exclusion*, on the other hand, is relatively straightforward and does not raise these statistical issues. When the profile of the specimen does not match the reference sample, the reference individual cannot be the source of the evidence sample. End of story. (In the case of trace DNA samples and mixtures, there can be exceptions to this simple rule, as we'll see in Chapters 4 and 5.)

In response to Ed's dilemma concerning the autopsy samples, I suggested that one approach might be to extract DNA from the evidence sample and carry out what was known as restriction fragment length polymorphism (RFLP) analysis, an approach using DNA gel electrophoresis and a method for analyzing the pattern of DNA fragments known as the Southern blot, named after Ed Southern, a distinguished Oxford biochemist, who developed the procedure.[1]

For genetic typing, RFLP analysis requires the extraction of a large amount of DNA from the sample; in addition, the DNA must consist of

---

[1] Inevitably, the propensity of science graduate students and postdocs for arcane self-amusement gave rise to Northern blots, based on RNA electrophoresis, followed a few years later by Western blots, based on antibody detection combined with protein electrophoresis. That Eastern blots have not yet emerged may testify to the improved sense of humor for the new generations of students— or perhaps to the exhaustion of additional molecular categories (lipids, carbohydrates) that can be separated by electrophoresis.

intact fragments of relatively large size (several thousands of the individual building blocks of DNA, called bases). RFLP analysis was already being used in basic genetic research, and a brilliant young geneticist from the University of Leicester, Alec Jeffreys, was just starting to apply a variant of RFLP analysis to forensics cases, an approach he termed "DNA fingerprinting" because it generated a complex pattern of DNA fragments that varied among different individuals and so could be used for individual identification. Unfortunately, the DNA we extracted from the formaldehyde-treated autopsy samples in the *Pestinikas* case was severely degraded, that is, the DNA fragments we obtained were far too small (around 200–300 bases) to carry out RFLP analysis.

After my initial disappointment at this result, I suggested to Ed that we should try our newly developed method of targeted DNA amplification, PCR (polymerase chain reaction). My colleagues and I had just published the first account of PCR amplification of a specific human gene, beta-globin, in a test tube using a DNA polymerase, an enzyme that copies DNA inside the cell, from the well-known laboratory bacterium *E. coli*, a few months earlier in the journal *Science* (Saiki et al., 1985). PCR synthesizes millions of copies of a target DNA sequence in a test tube, using an enzymatic process similar to DNA replication within the cell. Because the method used repeated cycles of DNA replication in which the amount of DNA doubled at every cycle (i.e., in 20 cycles the amount of targeted DNA would increase by 2 to the 20th power or 1 million-fold) and achieved an exponential accumulation of a specific targeted DNA fragment, my colleague, Kary Mullis, who came up with the idea, termed the method the polymerase chain reaction or PCR.[2]

Our initial demonstration had amplified the beta-globin gene from human genomic DNA for the diagnosis of sickle cell anemia (SCA), a disease of the red blood cells caused by mutations in both the maternal and paternal copies of the beta-globin gene.[3] After Randy Saiki, a young and very bright

---

[2] The basic idea of PCR is so simple, that once it's been reduced to practice, that is, shown to actually work, it seems obvious. As the conceptual artist, Sol Levitt pointed out, "successful ideas generally have the appearance of simplicity because they seem inevitable." Kary Mullis's enormously consequential idea of the PCR was recognized by the Nobel Committee with the Nobel Prize in Chemistry in 1991. Michal Smith, who developed a method for chemically synthesizing short pieces of DNA, known as oligonucleotides, was a corecipient of the 1991 Prize. Synthetic oligonucleotides play a critical role in the PCR process, which spawned a burgeoning industry responding to the huge demand for these oligonucleotides. Realizing the value of the technology, Dupont Corp took the "it's obvious" view and tried to have the PCR patent overturned. As inventors, Kary and I testified in front of a federal jury in San Francisco and the jury unanimously upheld the patent in 1991. PCR is described in more technical detail in the Appendix along with a brief history of its development.

[3] The beta-globin gene encodes part of the blood protein, hemoglobin, that transports oxygen. The first demonstration of the molecular basis for a genetic disease was the discovery in 1949 by Linus

technician in my lab, had successfully amplified the beta-globin gene, defining the narrow set of conditions under which Kary's idea might be made to work, we developed a system for amplifying and analyzing the DNA sequence of a gene involved in the immune system, the HLA-DQA1 locus, given my long-standing interest in immunogenetics.

Because the HLA-DQA1 gene is polymorphic (meaning that its DNA sequence varies in the human population), it could, in principle, be used for individual identification. So, we decided to see if the new DNA amplification method, when used to analyze this polymorphic gene, might be able to provide meaningful data on the problematic samples from the *Pestinikas* case. Our first attempt using the DNA polymerase from *E. coli* on the degraded DNA isolated from the formaldehyde-treated autopsy sample failed. My technician (at the time and for the following 33 years), Dory Bugawan, then tried using a newly isolated and more efficient DNA polymerase, Taq polymerase from a bacterium[4] that lives in the Yellowstone hot springs, into the PCR amplification protocol. To our delight, she was able to amplify the polymorphic HLA-DQA1 gene using this new enzyme. When we compared the results between the second autopsy samples and the reference sample from the initial autopsy, the DQA1 genotypes turned out to be identical. That meant that, within the limits of the test, the second autopsy samples "matched" the reference sample. Consequently, the second autopsy samples were presumed to be derived from Joseph Kly, as were the first, inconsistent with the prosecution's conjecture of a sample switch carried out to cover up a negligent homicide. Thrilled that we were able to finally get some meaningful results from these problematic samples with this novel technology, we went over the data with Ed and he went on to present it to the court.

Pauling of a physical change in the beta-globin protein in SCA. In 1957, Vernon Ingram showed that the change was a substitution of the amino acid valine for glutamic acid, resulting from a mutation in the beta-globin gene. In 1979, Y. W. Kan demonstrated that cutting the DNA with an enzyme (restriction endonuclease) that recognized a specific sequence associated with the SCA mutation could serve as a diagnostic test. Our original PCR diagnostic test in 1985, inspired by Kan's work, was a prenatal test for SCA based on sampling *fetal cells*. Now, over 35 years later, I am working on a noninvasive prenatal test for SCA based on analyzing *fetal DNA in the mother's blood* in the Children's Hospital Oakland Research Institute, the same building where, over 50 years ago, the Black Panthers organized a community SCA clinic.

[4] Taq polymerase had been isolated from the bacterium, *Thermus aquaticus*, a year earlier by my colleague David Gelfand. This enzyme, which functioned at high temperatures, transformed the way PCR could be carried out, making it more efficient, specific, and automatable (Saiki et al., 1988.) A description of this bacterium and the use of this enzyme in PCR is now part of the Yellowstone Park Rangers' educational commentary.

Doing anything for the first time is complicated. Doing something for the first time in a courtroom, like introducing a new form of evidence—is extremely complicated and requires strenuous justification. In the FSA report on the *Pestinikas* case, Ed Blake stated, "To the best of my knowledge, this case is the first attempt to apply DNA typing technology in a criminal investigation within the United States." His FSA report went on to acknowledge the novelty of the technique and to justify its application in this particular case:

> In doing so, we have applied certain procedures that are at the cutting edge of current molecular biology research. There are a number of reasons for selecting this case as a first test of DNA typing. First, conventional genetics typing procedures have either been unsuccessful (ABO typing) or cannot be exploited given the state of the evidence (all enzyme and protein genetic markers). There are a number of built in controls for the experiment, such as multiple tissue specimens and tissue specimens from both autopsies which are not in dispute. The analysis is not limited by sample size so that independent retesting is possible. The major defect in employing this technology is its lack of long-term use in forensic investigations in general, and the lack of significant controlled experience with formaldehyde fixed tissues in particular.

The HLA-DQalpha genotype that was identified in both the tissue specimens from the first and second autopsy was present in around 10% of the Caucasian population. (The current official name for the gene is HLA-DQA1, but the name of our PCR-based genotyping system was the HLA-DQalpha test.) Ed's report concluded, "these findings fail to reveal a genetic difference between any of the tissues from the first and second autopsy of Joseph Kly." Since these test results, requested by the prosecution, failed to support their hypothesis, the prosecutor tried to have the DNA typing evidence excluded. The defense objected and, given the understandable reluctance of the prosecution to attack the reliability of tests that they had requested, the judge allowed the introduction of the DNA evidence in the courtroom. (A year later, a second-generation HLA-DQalpha test with increased discrimination power found that the genotype common to the first and second autopsies was present in only 1% of the population, so that the *inclusion/match* was more informative: there was now a 1/100, rather than a 1/10 probability of a coincidental match.)

The lab report by FSA was duly entered as evidence in the *Pennsylvania vs. Pestinikas* case and presented to the jury. For Ed and me, the prospect that this new technology could be immediately valuable for forensics analysis was exhilarating. I was thrilled that our newly developed PCR-based genetic typing system had been able to productively fuse the two realms of scientific research and criminal investigation and solve this particular forensic dilemma, and I looked forward to working with Ed on many more forensics cases in the future. As it turned out, the tissue-switching hypothesis was not ultimately central to the prosecution's case and, after a lengthy trial, Walter and Helen Pestinikas were found guilty of murder in the third degree and of reckless endangerment because of their negligent treatment of Joseph Kly.

The mid-1980s were a major inflection point in the history of forensic analysis of crime scene evidence. In 1986, as we've just seen, our lab, in collaboration with Ed Blake, carried out DNA analysis on autopsy specimens in the *Pennsylvania vs. Pestinikas* negligent homicide case using the newly developed DNA amplification technology of PCR. The following year, Alec Jeffreys and his colleagues at Leicester University applied a different DNA technology, that of RFLP, to a rape/murder case in the United Kingdom. These two cases utilized two novel DNA technologies first revealed to the world in two scientific papers published some months earlier.

The two articles whose implications have transformed the field of molecular genetics as well as judicial systems around the world were both published in 1985. I've already mentioned the one in *Science* my colleagues and I published in December on PCR. Earlier that same year, Alec Jeffreys and his colleagues described in the journal *Nature* a method of analyzing complex patterns of DNA that varied in the human population and that could be used for individual identification and the term "DNA fingerprinting" was born (Jeffreys, Wilson, and Thein, 1985a, 1985b; Gill, Jeffreys, and Werret, 1985). The method analyzed DNA fragments that varied in length at many different regions of the genome, called the variable number tandem repeat (VNTR) loci, using the RFLP genotyping technique. The genetic variation was based on differing numbers of copies of a given repeated DNA sequence so the VNTR loci represented a *length* polymorphism rather than the *sequence* polymorphism present in the HLA genes we used for our initial analysis. In the RFLP system, the pattern of the polymorphic DNA fragments is analyzed by gel electrophoresis, a method for separating the fragments based on length. Although the initial method required a great deal of intact DNA (decidedly more DNA than is present in many forensic specimens) and the resulting complex pattern derived from many different parts (loci) of the genome could not be easily interpreted in statistical terms, the transformative idea and experimental demonstration that DNA analysis could be used for individual identification was launched.

The first application of forensic DNA analysis in a criminal case in the United Kingdom, the Colin Pitchfork case, made use of the Jeffrey's DNA fingerprinting (RFLP of VNTR loci) technique. The method had been previously used for individual identification in an 1985 immigration case (Jeffreys, Brookfield, and Semenoff, 1985). The melodramatic twists and turns of this milestone case, which involved the rape and murder of two young girls, the first in 1983 and the second in 1986, have been described in great detail in *The Blooding* by Joseph Wambaugh. Although the Pitchfork case is generally recognized as the pioneering application of DNA analysis in the identification and conviction of a murderer, it also illustrated the potential of the technology to *prevent* a wrongful conviction. Jeffreys initially compared RFLP profiles from semen samples from the two victims and concluded that they had been killed by the same man, but *not* by the prime suspect at the time, Richard Buckland, a local 17-year-old, who had admitted to the second murder under questioning. Jeffreys noted later that he had no doubt Buckland "would have been found guilty had it not been for DNA evidence"; this case was the first in a disturbingly long string of coerced and false confessions corrected by DNA analysis.

In an attempt to identify the true assailant, the UK's Forensic Science Service (FSS) carried out an unprecedented mass DNA screening. They conducted a 6-month-long investigation in which 5,000 local men were asked to provide either saliva or blood samples. No matches were found. Then, in 1987, a man who worked at a local bakery was overheard in a Leicester pub telling his mates that he had received a payment for providing a DNA sample in the name of a fellow worker, Colin Pitchfork. The conversation was reported to the police and a month later, Pitchfork was arrested. His RFLP profile matched that of the two semen stains and he pled guilty to the two rape/murders.

Although these initial two cases applied transformative DNA technologies, both of which could be used for individual identification, the two technologies were very different. RFLP analysis requires large amounts of intact DNA and is based on analyzing variation in the *length* of DNA fragments. In contrast, PCR can be used to amplify even trace amounts of DNA and was focused, initially, on analyzing variation in the DNA *sequence* of a single HLA gene. Now, all forensic tests involve PCR and analyze both length and sequence variation.

In the more than 35 years since the initial publication of these two papers and these two seminal cases, identifying the source of biological evidence samples found at the crime scene by DNA analysis has become a mainstay of contemporary police work and judicial proceedings, not to mention current TV fare. For many people, their primary contact with the field of genetics and DNA analysis comes either through the fictional forensic cases on TV or the real ones described in the newspapers. Given the pervasive presence of forensic DNA testing in the media (all DNA, all the time), it's hard to remember that its introduction was so recent. Since our lab's initial work applying PCR to the analysis of a few autopsy samples in *Pennsylvania vs. Pestinikas* in the summer of 1986, millions of samples have been analyzed and many thousands of cases have been resolved using what are basically the same DNA techniques. All this genetic testing has been made possible because PCR amplification can create millions of copies of a specific DNA sequence, allowing subsequent analysis from only a few initial copies—far too few to analyze—present in the forensic specimen.

The ability of PCR to amplify trace amounts of DNA has allowed the genetic profiling of individual hairs, minute amounts of saliva, blood, bone fragments, semen stains—essentially, any biological specimen from which DNA can be extracted, even the smallest amount. In our 1988 *Science* paper on PCR with the thermostable enzyme Taq polymerase, we had demonstrated that we could amplify and detect a single DNA molecule.[5] In the more than three decades since our initial application of PCR to forensics analysis with a single genetic marker (the HLA-DQA1 locus), many additional genetic markers have been incorporated in forensic genetic profiling. A "match" between the evidence and suspect at a single locus provides, using some relatively straightforward calculations, an estimate that this match could have occurred by chance. As the number of genetic loci analyzed increases, the

---

[5] This demonstration was statistical; diluting a DNA solution into multiple parallel PCRs such that most reactions have zero molecules (no product) indicates that the few positive reactions were initiated with a single molecule. A more direct demonstration was the amplification and detection of DNA from a cell with a single genome, such as sperm. My Cetus colleague, Norm Arnheim, had the insight that genotyping individual sperm cells could be the basis of a novel approach to genetic mapping in humans. Genetic, as opposed to physical, mapping involves analyzing genetic recombination that takes place between specific genetic loci. The longer the distance between two loci, the higher the number of recombination events between those loci. Instead of analyzing individual children in a family, the traditional strategy, one could now simply analyze individual sperm cells, eliminating the "middleman" (Boehnke et al., 1989).

probability of a chance match decreases dramatically. With the number of genetic markers currently in use, identifying the origin of evidence samples is now so convincing from a statistical perspective that, in many cases, a unique individual can be implicated as the source of the specimen.

While enormously powerful in terms of individual identification, DNA testing does have some important limitations in a criminal context, some of them more obvious than others. Genetic testing has the potential to identify the source of a specimen but does not *per se* establish guilt. The probative value of this identification will depend on context and on the nature of the specimen. If a suspect is identified as the (highly probable) source of a hair found at a murder scene, this result puts him, or at least his hair, at the scene of the crime but does not establish *when*. It certainly does not establish the suspect as the murderer. Blood and semen are, in principle, more critically related to a crime. If, for example, the source of the semen taken from a rape victim is the alleged rapist, then this identification clearly points to the likely guilt of the suspect. Even in this situation, however, alternative hypotheses can be posed to account for the match, such as the claim of consensual sex in a rape case (see *California vs. Mack* in Chapter 2). Similarly, it can be claimed (and has been famously argued in the OJ Simpson trial) that a blood stain found at a murder scene that is clearly derived from the suspect or the victim is, in fact, the result of planted evidence.

Another theoretical possibility, given the ability of PCR to analyze minute amounts of DNA, is that the crime scene evidence DNA sample whose genetic profile has been interpreted as implicating a particular individual could be the result of a secondary contaminating source of DNA. Although much beloved by defense attorneys as well as by screenwriters,[6] this "secondary transfer" scenario is extremely unlikely to occur. It did, however, occur in one frequently cited case. In 2013, Lukis Anderson was charged with murdering Raveesh Kumra, based on finding trace amounts of Anderson's DNA on the victim's body. However, Anderson had been drunk and unconscious at a local hospital at the time of the murder. As it turned out, the same two paramedics who treated Anderson for intoxication were at the Kumra murder scene a few hours later and presumably unwittingly transferred Anderson's DNA to the victim's fingernails. Anderson was released from jail after five months. Although this case is often cited as an example in the "DNA is not infallible"

---

[6] This "DNA transfer" scenario appears in a fictional case involving my favorite fictional detective, Harry (*né* Hieronymous) Bosch (*The Night Fire*, 2019) and has been used repeatedly in the TV series *Bosch* and *Bosch: Legacy*.

narrative, the DNA testing *per se* was correct. Nonetheless, it illustrates the potential pitfalls of trying to interpret trace amounts of DNA and is a valuable reminder that identifying the source of a crime scene DNA sample can be unrelated to the crime and to the issue of guilt.

When used appropriately, of course, the introduction of DNA testing into the courtroom has helped base convictions on objective, reliable, and statistically meaningful data, properties not always associated with other forensics techniques. In fact, it is DNA testing that has "raised the bar" and revealed some other forensic tests of long-standing usage to be subjective and potentially unreliable and, in doing so, has helped prevent the conviction of the innocent. In addition, DNA testing has allowed the verdict reversal of hundreds of the wrongfully convicted, including many on death row. The first such exoneration by DNA analysis occurred in the Gary Dotson case (*Illinois vs. Dotson*), where our lab was able to show that the DNA profile of the 11-year-old semen stain did *not* match Dotson's, convicted of rape 10 years earlier, but it *did* match the profile of the alleged victim's boyfriend (case details in Chapter 3). It took well over a year for the State of Illinois to overturn Dotson's conviction. Since this first DNA-based exoneration in 1988, 575 convictions have been overturned based on DNA evidence as of 2023 (National Registry of Exonerations, Michigan State University, College of Law), 21 of them in death penalty cases. The objective data generated by DNA technology and the committed efforts of lawyers and social justice activists all over the country have revealed a disturbing level of wrongful convictions in our criminal justice system. Wider use of the technology and continuing efforts on the part of both the defense and prosecution should, we can hope, reduce the occurrence of these tragic injustices.

As we have seen, the first use of DNA evidence in a US court was the PCR-based genotyping of a single gene in *Pennsylvania vs. Pestinikas*, but the results were not central to the verdict, which was guilty of murder in the third degree and of reckless endangerment. The first instance in which DNA evidence was used to obtain a conviction was the Tommy Lee Andrews case (1987) based on the RFLP analysis of a semen sample. Andrews, a serial rapist, had been arrested in several cases in the 1980s. Already serving a 22-year sentence for rape, Andrews was accused of raping and stabbing a 27-year old Orlando woman in her home on May 9, 1986. The victim identified Andrews as her assailant. RFLP analysis of DNA samples of a semen stain from the crime scene matched the profile of a blood sample that had been taken from Andrews after his arrest. Lifecodes, Inc., a small, commercial forensics lab in

Stamford, Connecticut, carried out the tests, and the prosecution claimed that the match between the RFLP profile of the semen and the suspect's blood meant that there was a 1/10 billion chance someone else would have the same profile and presented this astronomical number to the jury. He was convicted of rape, aggravated battery, and burglary on November 6, 1987.

Although the DNA evidence did not play a critical role in the conviction in the *Pennsylvania vs. Pestinikas case*, the prosecution's request for genetic testing and the defense's insistence that the results be presented in this milestone case were momentous. They initiated the long and contentious introduction of DNA evidence into the courtroom, a process that has transformed the US judicial system.

2

Deciding What the Jury Sees

DNA and Admissibility

All of us, my research colleagues and I as well as Ed Blake and his forensics coworkers, were elated with the genetic results in *Pennsylvania vs. Pestinikas* and the potential future for forensic applications of PCR genetic typing. As a research scientist, however, I was at the time rather naïve about courtroom proceedings and the institutional challenges there could be to the admissibility of evidence produced by new scientific techniques. My blissful ignorance of these issues had not been perturbed by the *Pestinikas* case because the prosecution had requested the analysis but the result favored the defense, so there had been no formal legal challenge by either side to the admissibility of the evidence.

In a lengthy and somewhat lurid account of this case in the summer 1987 issue of the *Scientific Sleuthing Newsletter*, James Starr, a forensics legal scholar and newsletter publisher, noted that "this case was less than ideal as a proving ground for the efficacy and legal admissibility of DNA typing. But, in other ways, the case was startling extraordinary"; he even compared the grisly account of events to the works of Edgar Allan Poe and Stephen King. Had the admissibility of the DNA evidence been formally challenged, *Pennsylvania vs. Pestinikas* might not have become the ground-breaking case that first introduced PCR genetic testing to the US courts. However, since the genetic typing of these autopsy samples was *not* subjected to an admissibility hearing, this first US DNA case did not serve as a *legal* precedent for the acceptance of the results of PCR genetic testing in the courts.

New technology is typically subjected to an "admissibility" hearing to determine whether the results can be presented to the jury. My first experience with this procedure was in *California vs. Mello* (1989). In this rape/murder case with two suspects, and in the many subsequent cases in which I was called to testify, I would see at firsthand how the adversarial nature of US court proceedings deals with new scientific evidence. It was not a pretty sight.

Prior to the admission of forensic evidence produced by relatively new scientific techniques, admissibility hearings (known as "Kelly-Frye hearings" in California) are held before the presiding judge to determine whether the evidence and the technology that generated it are sufficiently reliable to be presented to the jury. A challenge to the admissibility of evidence can be initiated by either the defense or the prosecution. Clearly, the admissibility hearing is a necessary step in ensuring a fair trial, and evaluating or vetting a novel technology is an admirable goal. Any new technology should be subjected to rigorous scrutiny, as should the performance of the labs using it to analyze critical forensic evidence.

The primary criterion for admissibility using California's Kelly-Frye standard is whether there is a consensus in the scientific community that the method in question is reliable. The standard of "general acceptance" as a prerequisite for admissibility was established in a federal appeals court case in 1923 in *Frye vs. United States*, a case involving lie detectors. In the *People vs. Kelly* (1976), the California Supreme Court affirmed the Frye rule requiring a consensus in the scientific community, noting that the primary advantage of the Frye rule "lies in its essentially conservative nature." The court ruling went on to say that "for a variety of reasons, Frye was deliberately intended to interpose substantial obstacles to the unrestrained admission of evidence based on new scientific principles."

Basing admissibility on "general acceptance" by the relevant scientific community has the virtue of not requiring a judge, who may not have the requisite background knowledge, to decide whether a particular novel technology is scientifically reliable but, instead, defers to a scientific consensus. Such a conservative standard, however, could in some cases delay the admissibility of critical data generated by a novel but reliable technique, results which could have been instrumental in exonerating or convicting the accused. A method that dependably produced valuable results but that was still unfamiliar to many scientists might well be rejected by this standard. A distinguished jurist in a lecture titled "Science and the Law: Ships Passing in the Night" described science as searching for the truth while the law valued stability and precedence above all. (Some philosophers and sociologists of science would undoubtedly disagree with this simple and uncritical view of science.) DNA in the courtroom represents just one aspect of this contentious but ultimately invaluable marriage of unlikely partners with somewhat disparate value systems.

In 1975 Congress created new federal rules of evidence in which the primary criterion for admissibility was whether the scientific information would assist the judge or jury in reaching a decision. This somewhat looser standard was revised in 1993 by the US Supreme Court, re-establishing the "general acceptance" criterion and adding requirements for peer review of the techniques and experimental results. A more recent legal standard based on a Supreme Court ruling in *Daubert vs. Merrill Dow Pharmaceutical* (1993) states that "the trial judge must ensure that any and all scientific testimony and evidence admitted is not only relevant, but reliable."

The existence of different standards in different states complicates the already complicated and contentious issue of admissibility. Nonetheless, these standards serve an essential function in the judicial system. In my experience as an expert witness, however, they worked better in theory than in practice in our adversarial legal system. Lawyers like, above all, to *win* . . . and so the process by which DNA evidence was evaluated in those early admissibility hearings was not an edifying spectacle. If the DNA evidence favored the defense (typically an exclusion), some prosecutors found this new DNA technology to be "premature" or "unreliable." If the DNA evidence favored the prosecution (typically an inclusion), the prosecutors managed to develop a degree of comfort with the technology while the defense attorneys would inevitably see serious flaws, arguing that it was unreliable and therefore obviously inadmissible. One defense expert witness, demonstrably clueless regarding the technology, referred to the first PCR genotyping test (the HLA-DQA1 typing test) as "junk science," although it was never clear by what criterion this conveniently dismissive judgment was reached. Both sides could always depend on the kindness of strangers—that is, on finding a few expert witnesses with the appropriate opinions willing to argue the merits of the technology in front of a judge who, typically, struggled to understand the basic concepts. Some of the genetics expert witnesses who testified against the technology objected to the genotyping method itself while others objected to the statistical analysis indicating the probability of a random match.

California vs. Mello

The first admissibility hearing of PCR genetic typing of forensic evidence (in this case, a semen stain) was in *California vs. Mello*, a rape/murder case of an elderly woman (William Mello's grandmother) in Riverside County in 1989. Two young men, William Mello and Lorenzo Modesto were suspected. The non-DNA evidence favoring guilt for the murder was strong; the genetic typing from the semen stain related only to the charge of rape. Prior to the Mello trial, the prosecution had made a deal with Modesto, a minor, and therefore ineligible for the death penalty, and had obtained a murder conviction. Mello was charged with *both* rape and murder. Under California law, if he was convicted on both counts, the prosecution could "go for the gold"—the death penalty. Unfortunately for this plan, the HLA-DQA1 genotype of the semen did not match that of Mello, an exclusion that indicated, as the defense argued, that Mello was not the source of the sperm and therefore innocent of the rape charge. The prosecution vigorously argued that this result was unreliable and should not be presented to the jury, based on the testimony of several scientific experts in the admissibility hearing.

I did not testify in the admissibility hearing for *California vs. Mello*, but Ed Blake did. I was shocked when I read the transcripts when Ed returned from Los Angeles and brought them over to the lab. The prosecutor, Villia Sherman, attempted to discredit Ed as a scientist, referring snidely to his publication record as "hardly a stellar performance." She later referred to Ed's testimony as "defensive," an understandable response to what were clearly personal attacks. Having now watched innumerable reruns of *Law and Order* since the *California vs. Mello* trial, I've come to understand that undermining the opposition expert witness is just part of the game, but I was young and green then. It was also dismaying to read the testimony of some geneticists and molecular biologists in this and subsequent trials opining that this new DNA genotyping method (PCR) was unreliable, comparing it unfavorably to the RFLP technology with which they were more familiar. In most cases, however, their testimony revealed considerable lack of understanding of the PCR-based genotyping technology used in the HLA-DQalpha test, illustrating that *lack* of familiarity can also "breed contempt." Ironically, RFLP technology would be phased out several years later in favor of the more robust PCR-based genotyping tests, now the norm.

Many expert witnesses who *had* used PCR in their research labs, of course, testified to its reliability, and their view prevailed in the vast majority of the admissibility hearings. Nonetheless, I found this process both frustrating and irritating as the claims of unreliability emerged as a standard ploy by either defense or prosecution time after time. By 1989, our lab had been carrying out both basic and applied research characterizing the use of PCR on a variety of samples for four years (with many peer-reviewed publications) and Ed Blake had been applying the PCR-based genotyping system we had developed to many different forensics specimens and on many different cases since the initial *Pestinikas* case in 1986. But the technology was still somewhat unfamiliar to many of the opposition expert witnesses, and much of the opposing testimony remained ill-informed. In retrospect, I suppose I was somewhat defensive about the technology we had developed as well as somewhat arrogant—"We know what we're doing and *you*, apparently, don't"—and definitely naïve about the way the judicial system works.

By 1989, PCR had been used in many different labs around the world for basic research as well as for some clinical diagnostic tests. However, some opposing experts argued that forensic specimens were fundamentally different from the "pristine" samples used in medical research and diagnostics. They expressed concerns, given the sensitivity of the amplification technology, about potential *contamination*, that is the possibility that the forensics specimen might have acquired additional material from the environment that would be amplified by PCR and confound the result. This is a valid concern, but the specter of contamination was often raised in the absence of any evidence for it in the actual genotyping data. In most cases, a hypothetically contaminated sample could be identified as having more than the two alleles expected for a single-source sample. This theoretical concern about possible contamination in the lab could be and was addressed, to some extent, by careful and rigorously controlled work procedures in the lab and by always running a "negative control," a sample with no DNA, that should, absent any contamination, yield a test result of *no* genotype. These procedures dealt with potential contamination in the lab but the possibility that the evidence samples could have been contaminated in the field or during collection and transfer to the lab remained.

Typically, expert witnesses for the defense introduced the concern about potential contamination, although the most likely potential error would be a false exclusion rather than a false inclusion. Also, many of the opposition critiques discussed *potential* problems that could complicate the results of PCR-based genetic testing while the actual data in the forensics reports they were opposing typically showed no evidence of these problems. Although I found this process frustrating, I recognized it was critical for the community to consider these potential problems and devise procedures to minimize their occurrence.

Many of the expert witnesses who testified against PCR in admissibility hearings were very forceful, if not always very knowledgeable, in

their objections. One of these witnesses, to my great surprise, was a distinguished genetics colleague, then at the University of California at Berkeley,
Mary Claire King, who testified in *California vs. Mello* that the PCR-HLA-
DQ-alpha test had an intrinsic defect based on her misunderstanding of data
generated by a post-doc in her own lab using another technique (RFLP). This
testimony, eminently useful to opponents of the PCR genetic typing technology, was cited in many subsequent admissibility hearings. It took several
years and a publication whose coauthors included a well-known forensic geneticist in her Department at UC Berkeley, George Sensabaugh, and King's
postdoctoral fellow, Kay Lichtenwalter, who had carried out the RFLP study,
to correct this misleading and mistaken testimony. Regrettably, King, who is
now at the University of Washington, did not join us as a coauthor. Testifying
in a subsequent case years later, she conceded that she had been mistaken
in that initial testimony, though she claimed that she had been misled in
some undefined way by scientific colleagues at Cetus. Her contributions to
the mapping of the first breast cancer gene, BRAC1, subsequently identified
by Mark Skolnick at the University of Utah, have been deservedly well
recognized with many scientific awards.

In the *California vs. Mello* case, the judge ultimately decided to allow the
DNA evidence to be presented to the jury. After several days of deliberation,
the jury, convinced by a wealth of non-DNA evidence, convicted Mello of
murder but, based on the HLA-DQA1 result, acquitted him of rape, making
him ineligible for the death penalty. He is currently serving a life sentence.
Focusing on Mello, the prosecution had viewed the HLA-DQA1 genotype
of the sperm as an *exclusion* and an inconvenient result that they chose to
oppose. However, the HLA-DQA1 genotype turned out to match the other
suspect, Modesto, with whom they'd already made a deal, suggesting that
Modesto was, in fact, the rapist.

It is conceivable that in a rape committed by two men the majority of
the sperm would be from the second rapist. But the fact remains that there
was no physical evidence tying Mello to the rape. Clearly, it is critical to review new technology rigorously, but excluding valid data can have critical
consequences. Had the judge decided to exclude the evidence provided by
this DNA method that was novel at the time and had the jury convicted
Mello of rape as well as murder, he might well have been sentenced to death.

One of the puzzling aspects of this case was why the prosecution was so
committed to having the PCR HLA-DQalpha genotype evidence ruled inadmissible. The prosecution had a very strong case without any need for DNA

evidence for murder and was, as it turned out, clearly able to obtain a murder conviction for Mello. Could it really just have been about the death penalty? Could it have been the "awkwardness" of explaining to the jury that the DNA genotype of the sperm did not match Mello but *did* match the suspect Modesto, with whom the prosecutors had already made a deal? Ed thought that the vehemence of the prosecutor's fight to exclude the DNA evidence in *California vs. Mello*, which was heard in Riverside County, was related to a case in Los Angeles County, *California vs. McSherry* (1989), a kidnap and rape case involving a six-year-old girl. In this case, Leonard McSherry, a known pedophile, had been convicted prior to the results of DNA analysis, based on compelling eyewitness testimony; McSherry was identified in a line-up by the girl as well as by her four-year-old brother.

The results of the PCR HLA-DQ alpha test that Ed performed on the sperm, however, proved inconvenient for the prosecution: they excluded McSherry as the source of the semen specimen taken from the child. The prosecutor, Brent Ferreira, assured me that he didn't care about this problematic DNA evidence and that McSherry would stay in jail. Following the initial analysis with the HLA-DQalpha test, we tested additional genetic markers, as we developed them, on the forensics sample and each additional test resulted in an exclusion of McSherry as the source. In 1992, McSherry's request for a new trial based on this exclusionary evidence was denied; the court cited the State's experts' concerns about the PCR technology and these experimental results.

I corresponded with several of the expert witnesses who had advised the court that our PCR test was unreliable, explaining how it worked and why the results were, in fact, reliable and excluded McSherry. Many of them posited a "worst case" scenario in which the specimen had been contaminated. As in many such cases, opposition experts invoked the *theoretical possibility* of contamination without looking at the actual data to see whether there was any evidence of contamination in the genotyping results. This "Boy Who Cried Wolf" aspect of some opposing testimony made me wonder how credible would objections be in response to a *real* genotyping problem.

Finally, in 2001, additional PCR tests using the newly developed, more discriminating commercial genotyping technology based on many different genetic markers (the short tandem repeat or STR markers, described in Chapter 5) available then, showed, once again, that McSherry was not the source of the sperm. These results helped identify the actual offender, who had been convicted "of a subsequent crime." In 2005, McSherry was finally exonerated, and he was compensated by the State for thirteen years of wrongful incarceration.

Ed was so outraged by what he characterized as the prosecution's "vigorous and vicious" attacks on the DNA evidence and by the testimony of some of the expert witnesses in *California vs. Mello* that he sent a letter to Rockne Harmon, the Alameda County District Attorney, and one of the very few attorneys on either side who had developed an impressive familiarity with the new DNA technology as well as with many of the scientists who practiced it. Ed delivered a biting critique of the existing adversarial approach to admissibility, pointing out that "the question of general acceptance in the relevant scientific community is a threshold question not dependent on the result in the instant case. A procedure which is generally accepted in the appropriate scientific community is just as accepted whether the issue is advanced by the defendant or the State independent of who benefits from the result. . . . Since the *Kelly/Frye* issue is not dependent on the result, the issue should be litigated without the advocates knowing the result. Thus, the adversarial confrontation will focus on principle rather than outcome."

Ed's argument that the general acceptance of a novel technology should be independent of whether the testing result favors one side or the other made sense to me. My experience with these early PCR cases convinced me that the adversarial system with dueling expert witnesses working for one side or the other in a particular case was not an ideal way to determine admissibility. It does, however, seem entirely appropriate and, indeed, crucial for the prosecution and defense to be able to vigorously challenge the DNA evidence that *is* presented to the jury to ensure that the DNA analysis of a specific specimen was performed correctly and that the presentation is not misleading.

California vs. Mack

The first case in which I testified in an admissibility hearing was *California vs. Mack* (1989). This was another rape/murder case and this time, the HLA-DQA1 genotype of the semen stain taken from the victim matched that of the suspect, Paul Mack. In 1987, the body of Karen W, a waitress at the Peppermill restaurant in Sacramento, was found in her car. Witnesses testified that she had been on her way to meet the defendant for a photo session. The defendant initially denied knowing the victim and, in the admissibility hearing, the defense strenuously argued that the DNA technology was unreliable. The defense noted that the FBI was not yet using PCR analysis and suggested that I had been misleading the court by referring to an article we had published in the distinguished science journal *Nature*; they triumphantly produced a copy of our published Letter (Higuchi et al., *DNA typing from single hairs*, 1988) on the amplification and genotyping of the HLA-DQA1 gene from individual hairs using the technique involved in this case, implying that it was simply a non-peer-reviewed piece of correspondence. I had to explain that, in *Nature*, short articles were called "Letters" while long articles were called, rather unimaginatively, "Articles," and that they were both rigorously peer-reviewed. A remarkable characteristic of some defense attorneys (and shared with more than a few prosecutors) which I witnessed in many subsequent trials is their capacity for generating an instant, intense, and performative sense of moral outrage. The defense team seemed to think that my supposed misrepresentation of the *Nature* publication was a crime virtually on a par with the rape/murder with which their client had been accused (and was later convicted).

The defense continued to ask me a series of questions, most in that time-honored formula, "*Isn't it true, Dr. Erlich, that* . . . [fill in the blank]." My favorite blank was "that Dr. King is planning to sue you?" to which I allowed that I was unaware of any such plans. Apparently, Mary Claire King, who had criticized the PCR test in *California vs. Mello*, decided she had better things to do, but her testimony in the Mello case continued to be used by defense attorneys in subsequent cases for several years.

In *California vs. Mack*, the judge ultimately ruled the DNA evidence inadmissible. The defense's argument that the FBI was not yet routinely using PCR technology apparently weighed heavily in the judge's decision. The impact of this point on a nonscientifically trained judge is understandable: if you're not really able to evaluate a new technology, why not just defer to an established and scientifically respected institution like the FBI? The problem here is that the issue of scientific validity is replaced by consideration of the schedules and practices of a large government lab and bureaucracy. Ironically, the defendant, who sat through the admissibility hearing, was apparently more convinced by the DNA evidence than was the judge and changed his testimony, conceding that he had, in fact, had sex with the victim, that the sperm was his—but that the sex had been *consensual*. He was subsequently convicted of murder and rape.

In this case, the judge's ruling prevented the jury from considering the DNA evidence, but they came to the correct conclusion nonetheless. In some of the subsequent trials in which the DNA evidence has been ruled inadmissible, however, the outcome could have been more problematic. An inadmissible exclusion could result in wrongful conviction and an inclusionary result that was ruled inadmissible could prevent the conviction of guilty defendant. The lengthy and contentious disputes about admissibility that result from our adversarial judicial system have, then, very real consequences: *data excluded can mean justice denied.*

Although the frustration I experienced in the course of these early admissibility hearings focused on the potential exclusion of reliable DNA evidence, I am assured by my forensic scientist colleagues that a significantly more serious problem is the admissibility of dubious evidence, such as certain interpretations of hair morphology and bitemarks. The solution to both of these problems may be a more rigorous scientific peer review on behalf of the court rather than the existing system of testimony by adversarial experts.

In courts of law, judgments about the validity of scientific techniques have been typically made by nonscientists. The judges who presided over these admissibility hearings with the ultimate responsibility of accepting or rejecting the evidence struggled mightily to understand the issues, as presented by fiercely partisan advocates and their expert witnesses. There were, of course, legitimate concerns to be discussed, in particular, the statistical issues concerning the significance of a "match." However, the notion that the "truth" about the validity of a new technology and the evidence it has generated can be best evaluated in an adversarial setting with the scientific experts testifying for one side or the other seemed, to a research scientist like me, misguided.

A system in which scientific experts for one side question scientific experts for the other side in front of the judge who has the responsibility to make the ultimate decision sounds as if it could work quite well. In the United States, the actual cross-examination of scientific experts is carried out by lawyers, who typically don't understand the science very well, and are therefore constrained, in some cases, to parroting the scientific judgments of *their* experts or challenging the credibility ("*how much are you being paid?*" . . . is a favorite) of the witness. At times, while on the witness stand, I felt as if I were in a one-on-one basketball game with the cross-examining lawyer posing a variety of *gotcha* questions to which I would have to respond clearly and calmly. These questions often were based on a convenient technical "misunderstanding" unrelated to any scientific issue. Articles published in the scientific journal *Proceedings of the National Academy of Sciences* (*PNAS*) have a page marking saying "advertisement" to minimize mailing costs. One defense attorney, in a rape/murder case of a 7-year-old in Washington state where I testified, attempted to have one of our *PNAS* articles, rigorously peer reviewed prior to publication, excluded from evidence on this basis. The judge, following a lengthy lunch recess in which he, presumably, looked into this patently bogus claim, rejected this argument.

Other legal systems, such as those in many European countries as well as in Australia, have a system for evaluating new technology in which the scientific experts act on behalf of the court rather than testifying for the prosecution or defense. It may be inevitable and necessary for the final judgment concerning admissibility to be made by a legal authority (i.e., a judge), but I think it would be vastly preferable for the judge to be presented with the opinions of scientific experts working for the court rather for the opposing advocates. In a *New York Times* article, "Experts Hired to Shed Light Can Leave U.S. Courts in Dark," Adam Liptak quotes Judge Denver D. Dillard of the Johnson County District Court in Iowa City in a 2005 decision, "The two sides have cancelled each other out." Judge Dillard rejected both experts' conclusions and complained that "no funding mechanism" was available for him to appoint an expert who would work for the court rather than the prosecution and defense. Liptak went on to quote John H. Langbein, a Yale law professor, who wrote in the *University of Chicago Law Review* that a European judge who visits the US court system experiences "something bordering on disbelief when he discovers that we extend the sphere of partisan control to the selection and preparation of experts."

In the United States, however, lawyers are reluctant to relinquish control of the scientific experts. "If I got myself an impartial witness," the trial lawyer Melvin Belli once said, "I'd think I was wasting my money." Some judges and lawyers have openly wondered whether "any scientific expert is truly neutral." This skepticism misses the point, in my view. Of course, scientific experts are not "neutral," and scientific discussions and disagreements are often heated, acrimonious, and intense. These scientific expert opinions and arguments about admissibility, however intense, are not in the service of either prosecution or defense and could be offered about the technology *per se* rather than the performance of an individual lab. If these disputes, discussions, and questioning of the evidence and the technology that generated it were carried out by scientific experts appointed by the courts, a process known in Australia as "hot-tubbing" (don't ask), I believe our justice system would benefit substantially.

The counterintuitive notion that the truth about a scientific technique would emerge best from a conflict between two adversarial advocates committed to "winning" is reminiscent of Adam Smith's Invisible Hand, whereby the greater good of society is somehow supposed to be best served by the individuals within that society pursuing their own self-interests. Since 1776, when *The Wealth of Nations* was first published, most modern capitalist societies, while recognizing that Smith's idea has some intrinsic merit, have decided that regulation can help *guide* such an Invisible Hand toward the social good. Smith, himself, unlike his latter-day acolytes, recognized that some regulation actually benefits the operation of the free market.

People vs. Castro: RFLP on Trial

In the first few years of forensic DNA testing by RFLP, neither the criteria for establishing a match nor the calculations of match probabilities were subject to rigorous challenge. The first real challenge to the introduction of DNA evidence came in 1989 in *People vs. Castro* in New York. Once again, the RFLP analysis was carried out by Lifecodes. Eric Lander, a mathematician and geneticist and, until recently, the director of MIT's prestigious Broad Institute and a Presidential Science advisor, was the expert witness for the defense. Richard Roberts, a Nobel laureate who discovered and studied the bacterial enzymes (restriction endonucleases) that cut DNA at specific sites and were used to carry out part of RFLP analysis, was the expert witness for the prosecution.

Jose Castro was accused of murdering his neighbor and her daughter. The prosecution maintained that RFLP analysis of a bloodstain on his watch revealed a match with the victim. Lander was highly critical of the quality of the RFLP data as well as the lack of standards in declaring a match. In a highly unusual development in admissibility challenges, Roberts, the prosecution expert witness, and Lander, the defense expert witness, met, discussed the RFLP data, and issued a scathing joint report to the Court.

After an extensive admissibility hearing, the court ruled that the DNA evidence could be used to demonstrate that the bloodstain did not come from Castro (an exclusion) but could *not* be used to show that it came from the victims (an inclusion). An *exclusion*, the apparent difference in DNA profiles between the evidence and a reference sample, is relatively straightforward to detect and to interpret. A match or *inclusion* on the other hand, is more complicated. The criteria necessary to determine that DNA profiles from two different samples are actually identical must be established as do the appropriate procedures for calculating the probability of a coincidental match (the RMP). The joint report by Lander and Roberts was critical of the Lifecodes data and procedures on both scores.

The court's ruling in this case proved to be critical in all subsequent US admissibility hearings on DNA evidence. While finding that DNA forensics techniques were "generally accepted" by the scientific community, it stated that pretrial hearings should be required to evaluate whether the testing laboratory's results were consistent with scientific standards. The court also recommended extensive requirements for discovery of laboratory data and protocols for future admissibility hearings. Castro was convicted of murder, but the challenge to the presentation of the RFLP evidence in this case marked the beginning of a concerted effort by the government and the forensic community to establish standards and guidelines for all DNA forensic applications.

The "DNA Wars"

The Castro case and the associated challenges raised in the admissibility hearing represented a pivotal moment in the history of DNA evidence in the courtroom and led to a period of dueling scientific publications known as the "DNA Wars." The battle started with a 1989 editorial in the highly respected journal *Nature*, titled "DNA Fingerprinting on Trial" by Eric Lander (Lander, 1989), based on his experience as a defense expert witness in *People vs. Castro*. He focused on the lack of standards in the application of RFLP analysis for forensics specimens. He also challenged the assumption of statistical independence of the various probability estimates for the individual genetic markers.

Two years later, in 1991, two well-known and respected population geneticists, Dan Hartl and his Harvard colleague, Richard Lewontin, wrote a blistering critique in *Nature* of the forensic use of DNA analysis (Lewontin and Hartl,1991). While they were highly critical of many aspects of forensic DNA analysis, their main criticism focused on the assumptions made in the statistical estimates of RMPs, in particular, the use of the Product Rule and the assumption of statistical independence that allows the probabilities of the individual genetic markers to be multiplied. Like Lander, they argued that available population genetics data for RFLP profiles did not support the simplifying assumption that the frequency of a given DNA profile at one genetic locus was statistically independent of the frequency at other loci. Lewontin, as we'll see in Chapter 13, made major contributions to our understanding of how genetic diversity is distributed in human population well before the DNA era. He was an elegant writer and brilliant polemicist whose articles often appeared in the *New York Review of Books* and someone who clearly relished a good fight. His contrarian spirit and objection to modern genomics as overly reductionist is captured by the title of an essay collection, *It Ain't Necessarily So: Dreams of the Human Genome and Other Illusions* (2001). His political tendency to poke and provoke is reflected in his statement that DNA is the ruling class while protein is the working class. While I disagreed with his conclusions about forensic genetics and the Human Genome Project, I had enormous respect for his research and intellectual flair.

Two other distinguished geneticists, Ranajit Chakraborty and Kenneth Kidd responded in the pages of *Nature* (Chakraborty and Kidd, 1991), vigorously defending the interpretation of the population genetics data and the methods used to estimate RMPs. The following year, they were joined by another well-known statistical geneticist, Bruce Weir, who entered the fray (Weir, 1992) arguing that the assumption of statistical independence was, in fact, consistent with the genotype frequency data in population databases.

A truce in the battle over the appropriate interpretation of population genetics data and the statistical estimates of RMPs was established with the publication of "DNA Fingerprinting Dispute Laid to Rest" in *Nature* (Lander and Budowle, 1994) coauthored by Bruce Budowle of the FBI and Eric Lander, who had fired the first shot in the DNA Wars in 1989. The paper presented a "new consensus" on DNA evidence and represented a cease-fire leading to a negotiated peace, as did the guidelines set forth in the 1992 National Research Council Report, *DNA Technology in Forensic Science*. One of the developments that helped end the early phase of the "DNA Wars" was the establishment, by the FBI and other members of the forensic community (the Scientific Working Group on DNA Analysis Methods or SWGDAM), of validation guidelines, specific experiments and procedures to be carried out for any new forensic genetic technique to be accepted in the courtroom. The 1992 commission, chaired by the medical geneticist Victor McKusick, addressed but did not really resolve some of the contentious issues surrounding DNA in the courtroom. A subsequent commission in 1996, chaired by the brilliant population geneticist James Crow, then around 75 years old, resolved many of the statistical controversies surrounding forensics DNA analysis in *The Evaluation of Forensic DNA Evidence*, a report so clear that one could learn the basics of population genetics theory from a careful reading of its text. I had the opportunity to serve on another committee chaired by Crow (National Research Council, Committee on DNA Technology in Forensic Science, 1996–97) and it was a great pleasure to see Crow, a gracious and charming chairman, apply his rigorous logic and lucid writing to what had been a disputed and contentious area.

Although these early years of DNA evidence in the courtroom were highly contentious and for me, often very frustrating, the general acceptance of PCR-based DNA analysis for single-source samples has been well established for many years. Some issues of admissibility remain for analysis of mixed samples, and we can anticipate challenges to the admissibility of evidence generated by recent developments in DNA technology such as Next Generation Sequencing and Rapid DNA (see Chapter 5) but the DNA Wars are over . . . and a battle-scarred and wiser DNA technology emerged from the fray.

3

Exclusions and Exonerations

Justice for the Wrongfully Convicted

While Ed and I were very excited about the potential of this new technology for analyzing difficult forensic specimens, we both were aware that, with our initial PCR-based test based on only a single gene, the significance of a "match" was still limited, not much more informative than an ABO blood group type. An "exclusion," however, was a different story. If the genetic profile of the crime scene evidence does not match the suspect, the interpretation is straightforward. The source of the evidence sample is *not* the individual who provided the reference sample, and no complicated statistics need be invoked. One of the striking—and sobering—observations we made during our first 200 cases was that in about 35% of cases, all of the suspects in a given case were excluded by our initial PCR genetic test, the HLA-DQalpha test (Blake et al., 1992). A similarly high rate of exclusion was seen in crime labs in the United Kingdom and United States using restriction fragment length polymorphism (RFLP) analysis. In some cases, the DNA result might be accounted for by supposing that the evidence was irrelevant to the crime and that the suspect was, in fact, guilty. In most of these cases, the state declined to prosecute the excluded individual. The inescapable conclusion, however, was that innocent people were being charged in a significant proportion of these cases. One would hope, of course, that, in the absence of DNA evidence, the vast majority of these innocent suspects would still have been found not guilty. But we now know, thanks to the national Innocence Project, founded in 1992 by Peter Neufeld and Barry Scheck as a law school clinic at Cardoza Law School and to the many state Innocence Projects, that a disturbingly large number of innocent suspects are nonetheless convicted. DNA testing could, in principle, eliminate these wrongful convictions whenever probative DNA evidence is available and considered at the trial.

DNA testing could not only *prevent* such convictions, but given the ability of PCR to amplify genetic markers from the degraded DNA found in old evidence specimens, we thought it had the potential to allow the review of old cases and, potentially, exonerate the wrongfully convicted.

The first such exoneration by DNA analysis was the Gary Dotson case, based on the results of our HLA-DQ alpha test. Dotson had been convicted in Illinois of aggravated kidnapping and rape in 1979, based on the victim's testimony and analysis of pubic hair. In 1985, Cathleen Webb, only 16 years old at the time of the alleged rape in 1977, had recanted following a religious conversion, explaining that her fabrication was intended to hide having had sex with her boyfriend. The prosecution, however, was unwilling to overturn Dotson's conviction and in 1987, the Appellate Court of Illinois affirmed the conviction and Gary Dotson continued to languish in jail (or on parole). In 1988, Dotson's new attorney sent the 11-year-old semen stain obtained from the (alleged) victim to Alec Jeffreys, the pioneering geneticist who had worked on the Colin Pitchfork case (see Chapter 1) Jeffreys applied RFLP analysis to these samples but failed to obtain a result. As in the *Pennsylvania vs. Pestinikas* case, the DNA was too degraded for RFLP.

In addition to the false testimony, misleading forensic results on blood typing and hair morphology had also contributed to Dotson's conviction. A pubic hair found in the rape kit was said by the forensic analyst to be "similar" to Dotson's. The analysis of hair morphology is notoriously subjective, and the significance of an alleged "match" is neither clear nor reliable. While the blood types A and B were found in the rape kit, the alleged victim and Dotson were both B, raising the question, where did the blood group A come from? In spite of the admittedly false testimony and the questionable forensic results, Dotson's conviction was upheld. Many prosecutors are notably reluctant to overturn convictions, particularly when the convicted suspect is, in their view, a less than exemplary citizen. Apparently, Dotson did not meet this high standard of citizenship.

Shortly after the failure to obtain an RFLP result, Mark Stolorow, a forensic scientist from the Illinois State Police, brought the semen stains from this case to our lab and Forensic Science Associates (FSA) so we could try our new PCR-based test. When we completed the analysis of the various semen stains and the relevant reference samples, the results supported Webb's recantation. The HLA-DQalpha genotype did not match that of Gary Dotson, an *exclusion*, but *did* match that of her boyfriend, David Beirne. Even after this compelling evidence of Dotson's innocence, the prosecution was still reluctant to accept these results and to exonerate Dotson. We sent the results of our HLA-DQalpha analysis to Jeffreys, and he wrote two letters on Dotson's behalf to

Dotson's lawyer, T.M. Breen, concerning the PCR data. In one, dated April 18, 1989, Jeffreys concludes that the PCR analysis "provided convincing evidence that the semen recovered from the knickers was not from Dotson and had genetic characters consistent with the semen having come from David Beirne." A few days later, following a phone discussion with Stolorow and the receipt of additional PCR DQ-alpha test results on several different DNA extracts, including one that Jeffreys himself had prepared from the semen stains, he sent another letter. He concludes the letter:

> In view of the conclusive nature of this evidence, I earnestly hope that you will be able to go back to the court on this matter and obtain the release of your client. It is clear that the rejection of this evidence by the judiciary would constitute a gross miscarriage of justice.

Presumably, Jeffrey's letters had an impact. Over a year and half after the initial FSA report had been submitted, the State of Illinois finally overturned the conviction in 1989 and Gary Dotson became the first in what is now a long list of wrongly convicted people who have had their convictions overturned by DNA evidence. Some prosecutors are eager to have wrongful convictions overturned; others, as in this case, seem reluctant to do so, even in the face of clear DNA evidence. What emerged from this and subsequent cases, is that the exclusion by DNA testing of a previously convicted and incarcerated individual, was often the just the *beginning* of a slow and lengthy exoneration process. Since this initial DNA-based exoneration in 1989, 575 convictions have been overturned based on DNA evidence as of 2023 (National Registry of Exonerations, Michigan State University, College of Law), 21 of them in death penalty cases. An example of an exclusionary result with the HLA-DQalpha test, the first PCR-based genetic test kit,[1] is shown in Figure 3.1.

The Dotson case involved a very rare example of *false* eyewitness testimony by the victim. *Incorrect* eyewitness accounts are much more common. Many of the other early FSA cases of postconviction review revealed that, in the trauma and tumult of violent crimes like rape, eyewitness accounts by victims could be unreliable. Glenn Woodall was accused of rape, kidnapping,

---

[1] We later applied the same technology, used here to analyze genetic variation in the HLA-DQA1 gene, to develop a test for analyzing genetic variation in the human papilloma virus (HPV) that causes cervical cancer. As it turned out, only some of the variants of HPV, for example, HPV16, confer high risk for cervical cancer. Most of the current HPV DNA tests that are now replacing the Pap Smear as a screening test distinguish the high-risk HPV variants from their low-risk brethren.

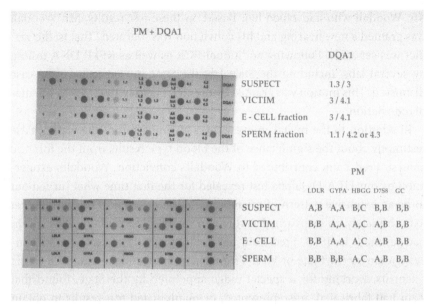

Figure 3.1 The blue dots indicate where the amplified DNA has bound to specific immobilized probes, indicating the genotype for the HLA-DQA1 gene and for the five co-amplified genetic markers. The combined test was known as the Polymarker test, launched in 1991. The semen stain was separated into the sperm fraction and the epithelial cell fraction. In this case, the genotype of the e-cell fraction matches the victim, as expected. The genotype of the sperm fraction does not match that of the suspect, resulting in an exclusion.

and aggravated assault in two separate incidents in West Virginia. The testimony of one of the victims had helped convict Mr. Woodall. Conventional serology and hair analysis had not excluded him. DNA testing was not performed, although requested by the defense, and Woodall was sentenced to life without parole in 1987. After the trial, RFLP testing was carried out but the results were inconclusive. Woodall continued to request additional DNA testing and, eventually, the PCR-based HLA-DQalpha testing was performed by FSA on sperm samples from the rape kits of both assaults. The HLA-DQalpha types from both samples matched, indicating a single rapist responsible for both assaults. They also excluded Woodall. Even after the results of the PCR HLA-DQalpha test showed that the semen stain in the rape kits did not match Mr. Woodall's genetic profile, the victim who had testified in the trial maintained her belief and contention that it was, in fact,

Mr. Woodall who had raped her. Based on these test results, Mr. Woodall was granted a new hearing and his conviction was "vacated," that is, the verdict was set aside. Following additional PCR as well as RFLP DNA testing by several labs, including the State lab, the State moved to have the case dismissed. This motion was granted by the court in May 1992, five years after his conviction.

In addition to the mistaken eyewitness testimony from the victim, false testimony about the significance of the blood type results from the forensic analyst, Fred Zain, contributed to Woodall's conviction. Woodall's exoneration by our HLA-DQalpha test revealed for the first time what turned out to be an extensive pattern of misrepresentation of forensic test results that led, ultimately, to overturned verdicts in many other cases in which Zain had testified. In 1993, Fred Zain became the subject of a misconduct investigation by the State of West Virginia. An independent team of forensic scientists, working for a special judge appointed by the State, found that Zain had fabricated, misrepresented, or manipulated test results to obtain convictions in all of the 36 cases under review. Zain died of cancer while awaiting trial. Misrepresentation of the laboratory results by a forensic examiner is, of course, a serious crime. It is also an egregious violation of a professional code. As the forensic scientist, Robin Cotton, said in a panel discussion, "The prosecutor speaks for the victim, the defense attorney speaks for the defendant, and the analyst speaks for the data."

The wrongful conviction in the Dotson case rested on *false* testimony by the victim (alleged) and in the Woodall case, on *mistaken* victim testimony. A more common problem is *false* testimony by a critical witness. In 2022, a judge, overturned the conviction of George De Jesus and his brother, Melvin DeJesus, in the 1995 rape/murder of their neighbor, Margaret Midkiff in Pontiac, Michigan (Levenson, 2022). Although the brothers maintained their innocence, they were convicted and sentenced to life without parole in 1997, based on the testimony of Brandon Gohagen. The brothers' claim of innocence was supported by the Cooley Innocence Project at Western Michigan University and the University of Michigan and, in 2020, the case was reviewed by the Michigan attorney general's Conviction Integrity Unit. DNA analysis revealed that DNA on the victim's body was consistent with Mr. Gohagen's DNA but excluded both of the brothers. The DNA results, however, were not inconsistent with Mr. Gohagen's testimony at the trial, which claimed that the brothers had forced him to rape Ms. Midkiff but that they had committed the murder. The review by the Conviction Integrity Unit found that Gohagen testified against the DeJesus brothers in exchange for a deal that allowed him to plead guilty to lesser charges and avoid a mandatory life sentence. In 2016, DNA evidence in another rape/murder case in Pontiac cast doubt on Gohagen's testimony and began to undermine the case against the brothers. In 1994, Rosalia Brantley was sexually assaulted and violently killed 11 months before the Midkiff murder and only three miles from the Midkiff crime scene. The details of both crimes were strikingly similar, and the DNA evidence from the Brantley case also implicated Gohagen as the assailant. He was convicted and sentenced to life without parole in 2017. This DNA evidence, as well as the testimony of 12 other victims of physical and sexual abuse, destroyed any remaining credibility of Gohagen's initial testimony in which he claimed that he had been forced by the DeJesus brothers to rape Ms. Midkiff. So, 25 years after the DeJesus brothers had been sentenced and 35 years after the initial DNA exoneration of Gary Dotson, DNA evidence in another rape/murder case resulted in the overturning their conviction, which had been based on the false testimony of the true perpetrator.

One of the recurring themes of DNA exoneration cases is that the State is often reluctant to overturn a conviction until the true perpetrator has been identified. It's understandable that prosecutors might experience a "need for closure" but justice should not be denied or delayed while the search for the actual perpetrator continues. Wrongful convictions need to be overturned as soon as evidence has shown them to be wrong.

In 1990, Jeffrey Mark Deskovic, aged 16, was convicted of raping and strangling a Peekskill High School classmate, Angela Correa, in Westchester County, New York. Although the DNA evidence present at his trial showed that the semen found in the victim's body *could not have been his*, the police presented a confession obtained after six hours of questioning. They also provided a theory to explain the exclusionary result, namely that the semen came from consensual sex with someone else. Apparently, the jury found the confession more persuasive (or understandable) than the results of DNA analysis, which was still quite new at the time. They may also have believed the prosecutor's theory of consensual sex with "Some Other Dude." This theory might have been more compelling had the prosecution actually found a boyfriend who did match the semen DNA profile . . . but they did not. This case illustrates dramatically the unreliability of confessions, a form of evidence that many of us had assumed (naively, as it turned out) to be definitive. At the news conference following Deskovic's release from prison, Barry Scheck noted, "We've learned a lot about false confessions in the past

decade. Videotaping of confessions and training of police officers could definitely lead to different results" (Santos, 2006).

As with many exonerations, Deskovic's conviction for rape and murder
was not overturned until a database search identified the source of the semen
specimen and Correa's actual assailant. After years of legal battles, Scheck
was able to convince the Westchester County district attorney, Janet DiFiore,
to compare the genetic profile of the semen specimen to a national data-
bank of convicted criminals. The semen DNA profile turned out to match
the profile of a man currently serving time for another murder, who then
confessed to having raped and murdered Angela Correa. The painful irony
of this case is that exculpatory DNA evidence was available at the 1990 trial
and it took a database search to reveal the false nature of the confession. In
2014, Deskovic was awarded $40 million by a federal jury after finding that
Daniel Stephens, the former sheriff's investigator, had fabricated evidence
and coerced Deskovic's false confession.

Financial compensation is, sadly, not always available to the innocent
people who have languished in jail due to a wrongful conviction. Sometimes,
even after it *has* been awarded, the payment is refused. In 1995, Darryl
Howard was convicted of the 1991 murders of Doris Washington, 29, and
her daughter, Nishonda, 13, and sentenced to 80 years in prison. DNA testing
from the semen found in the victim's body, *excluded* Howard, but the Durham
County prosecutor, Mike Nifong, nonetheless charged Howard with sexual
assault and murder. Nifong and Durham Police Detective Darryll Dowdy
argued that the semen came from a boyfriend, although they did not pursue
this speculation.

On August 31, 2016, a North Carolina judge vacated the conviction,
after finding police and prosecutorial misconduct, including the coercion
of false testimony, later recanted by the witnesses. Governor Roy Cooper
granted Howard a pardon of innocence the same day. Howard's lawyers
had presented new evidence that the DNA profile of the semen, which had
excluded Howard, linked two other people to the 1991 crime scene. In 2021,
in response to a civil rights lawsuit, against Dowdy and the city of Durham, a
federal jury found that Dowdy had fabricated evidence and awarded Howard
$6 million. On April 22, 2022, however, the city of Durham, refused to pay
Howard, arguing that Dowdy acted in "bad faith" and therefore declined to
indemnify him from the jury award. And so the search for justice continues.
Of the 375 DNA exonerees since the Dotson case in 1989, only 268 have been
compensated (Innocence Project, 2018).

In most of these exonerations, DNA evidence revealed the innocence of someone convicted in the *absence* of DNA analysis. As Peter Neufeld observed, "When you hold DNA up to the mirror and that mirror is the criminal justice system, the DNA points out all the cracks in that mirror in a way that we never saw before" ("What Jennifer Saw," 1997). In some cases, as we've just seen, the exclusionary DNA evidence was ignored. In a few cases, the conviction was based on DNA results that were misinterpreted. Or worse, misrepresented.

Every one of these 375 exonerations has its own harrowing story of injustice and personal suffering, but the most dramatic postconviction review case in which I was involved was *Virginia vs. Earl Washington, Jr.* in 1994. This death penalty case, like many others, raised serious issues about the reliability of some confessions, particularly those of vulnerable suspects. It also illustrated the difficulty in interpreting the DNA profiles obtained from forensics mixtures. My report analyzing the data, which argued for the suspect's innocence, was submitted just days before his scheduled execution. This case deserves its own chapter (Chapter 4).

4

The Earl Washington Jr. Case and the Problem of Mixtures

Earl Washington Jr. was convicted of rape and murder, based on his confession and the prosecution's interpretation of the HLA-DQalpha results, and sentenced to death. In my review of the data, I interpreted the results of the DQalpha test on the vaginal swab and crime scene samples as an *exclusion* and submitted my report a few days before the scheduled execution. This case, like so many others, raised serious and disturbing issues about the reliability of some confessions. It also highlights the difficulty in interpreting forensics mixtures.

In 1982, Rebecca Williams, a 19-year-old girl was raped and brutally stabbed to death in her apartment in Culpeper, Virginia. Before she died, she identified her attacker as a black man. A year later, Earl Washington Jr., an intellectually disabled 22-year-old African American man, was arrested and charged with burglary and assault in a nearby county. The police accused Washington of Williams's rape and murder and, following interrogation, he confessed. At his trial, however, Washington maintained his innocence but he was convicted and sentenced to death by a Virginia court in 1984. His execution was scheduled for January 1994. Washington's attorney, Barry Weinstein, was convinced of his innocence and, a scant 10 days before his scheduled execution, asked me to review the results of the PCR HLA-DQalpha testing of the semen stain and vaginal swab sample carried out the by Virginia state lab.

The DNA analyst in the Virginia lab, Jeff Ban, performed the test on a DNA from a vaginal swab taken during the autopsy on Williams as well as on a semen stain on a blanket. Washington was excluded from the semen stain but the vaginal swab specimen was a mixture, with DNA from more than one contributor, a notoriously difficult kind of forensic sample to analyze and interpret. Ban reported that, although the test identified a genetic variant (allele) absent in Washington, the victim, or the victim's husband, he *could not eliminate* Washington as a potential source of the semen. The prosecution argued, on the basis of the lab report, that the DNA results indicated an *inclusion*, consistent with the guilty verdict, because his *alleles* were present in the mixture. My interpretation of this complex mixture, however, was that Earl Washington's Jr.'s *genotype* was *not* present in the mixture (an *exclusion*) and, therefore, that he was likely innocent of the charges for which he'd been convicted.

The execution date was just days before Governor Douglas Wilder, the first African American governor of Virginia, was scheduled to leave office. The defense hoped that Wilder, a Democrat, might be more sympathetic to a plea on Washington's behalf than would his successor, a law and order Republican, George Allen. I received the lab report by Federal Express in the afternoon, stayed up most of the night evaluating the data, and faxed in my own report the following morning of January 13. On January 14, Governor Wilder commuted Washington's sentence to life in prison but declined to pardon him.

The test results on the vaginal swab taken from the victim were not at all straightforward, as is typically the case with a mixed sample. There was clear evidence that at least two different individuals had contributed to the sample; this *mixture* was inferred from the presence of three different HLA-DQalpha alleles. For any given gene, a single individual has two alleles, one inherited from the father and one from the mother. Somatic cells, all cells other than gametes, like sperm and eggs, are all *diploid*, that is, contain two copies of each autosomal chromosome, (i.e., not the X or Y chromosome) and of each gene on those chromosomes. (Red blood cells, which don't have a nucleus, are the exception.) Gametes, on the other hand, are *haploid*, containing only one copy. If more than two alleles are present in the DNA derived from the forensic specimen, then more than one individual has contributed to the sample. These forensic mixtures represent a technical challenge because it is often difficult to interpret the data as a mixture of two or more *specific* genotypes. The Virginia DNA lab report stated that three different HLA-DQalpha alleles were identified in the vaginal swab: 1.1, 1.2, and 4. The reference sample for Mr. Washington was a 1.2/4 genotype and for the victim, a 4/4 genotype. The conclusion of the lab report was that, since the mixture contained both a 1.2 and a 4 allele, the analysis of the swab was compatible with Mr. Washington's being one of the contributors to the sample, ignoring the unexplained presence of the 1.1 allele, absent from both the victim and Mr. Washington. The report concluded that, therefore, the DNA evidence provided no basis on which to reconsider his conviction and imminent execution.

But alleles do not exist independently in nature; they occur in pairs, as genotypes, in diploid somatic cells. In the case of sperm (haploid) from a 1.2/4 individual, one would expect an equal amount of 1.2-bearing and 4-bearing sperm. Consider a mixture of these three alleles in which the alleles were present in equivalent amounts. This situation would pose major difficulties in interpreting the data since this result is consistent with a mixture of many possible genotypes. Had this been the case, I would have been unable to exclude Earl Washington Jr.

In the HLA-DQalpha test, the sequence of the PCR-amplified DNA from the sample could be inferred by observing whether the amplified DNA bound to a panel of different short synthetic DNA probes corresponding to the sequence variants of this gene. If the DNA amplified from the evidence sample bound to a specific probe, a blue colored dot was generated in a specific position. The intensity of the blue color indicated the amount of amplified DNA that had bound to the probe, as illustrated in Figure 3.1 of Chapter 3.

In the case of the HLA-DQalpha results in the Washington report, the intensity of the blue dots identifying the 1.1 and 1.2 alleles was strong while the intensity of the blue dot for the 4 allele was much weaker. In interpreting this mixture, I made two critical assumptions. First, given my familiarity with this assay, I assumed, based on the intensity of the blue dots, that the two DQalpha alleles 1.1 and 1.2 constituted the majority genotype (1.1/1.2) and there was a trace amount of a third allele DQalpha 4. Second, since the separation of sperm from a semen stain or vaginal swab often results in residual trace amounts of the victim's epithelial cells, I assumed that the minority component in this mixture was a 4/4 contribution from the victim. Alternatively, but less likely, the minor DQalpha 4 allele could have come from the husband. This analysis implicated an individual with the DQalpha 1.1/1.2 genotype as the source of the sperm; this interpretation excluded Earl Washington Jr., who had the 1.2/4 genotype.[1]

Interpreting mixtures is always hard. It is even harder when the analysis is based only on a single gene. I was able to infer that Washington's HLA-DQalpha genotype was absent from the semen stain *only* because the amounts of the two components of the mixture (rapist and victim) were so

---

[1] An alternative, if unlikely, scenario, consistent with the DQalpha results, was that the semen sample mixture was a mixture of a majority genotype 1.1/1.2 and a minority genotype 1.2/4. This interpretation *would include* Washington but would require an altered prosecution "theory" (more precisely, "narrative") of *two* rapists, inconsistent with any of the testimony in this case. Clearly, the exclusionary interpretation was much more likely.

different. If the 1.2 allele and the 4 allele blue dots had similar intensities, these two alleles (1.2 and 4) *could* have belonged to the same individual, matching Earl Washington Jr.'s genotype, and the State of Virginia would have proceeded with the scheduled execution. Fortunately, DNA technology has evolved and there are now newer and much more powerful ways of analyzing mixtures than our original HLA-DQalpha test. In spite of the HLA-DQalpha evidence and my report, Washington remained in jail. In 2000, following significant advances in forensic DNA technology and the continuing doubts about Washington's guilt, then Governor Jim Gilmore, who would briefly run as a presidential candidate in the crowded 2016 race for the Republican nomination, ordered the Virginia lab to perform additional testing with the current STR (short tandem repeat) genotyping systems (discussed in Chapter 5). Ban's testing of the semen stain on the blanket, once again, excluded Washington. A comparison of the genetic profile obtained with STR technology with the profiles in the state's criminal DNA database, yielded a match with Kenneth Tinsley, a convicted rapist. However, Ban's analysis of slides from the vaginal swab was puzzling; his results excluded both Washington and Tinsley and, surprisingly, identified two additional genetic profiles.

Based on the exclusionary results for the semen stain and the vaginal swab, Gilmore granted Earl Washington Jr. a pardon. The pardon, however, was limited and Washington was released from prison to parole supervision in 2001. Since Ban's puzzling results on the vaginal swab slides excluded Tinsley, the prosecution declined to charge Tinsley and the Williams case remained open. Fortunately, duplicate slides were sent to Ed Blake. Using the same STR technology, Ed's results identified Tinsley's genetic profile in the mixture and excluded Washington. The serious and disturbing problems with Ban's results revealed in this case led to an independent investigation of the Virginia state lab ordered by then Governor Mark Warner.[2] To paraphrase the NRA, the best response to a bad lab with DNA—is a good lab with DNA.

In 2007, Kenneth Tinsley, convicted of rape in Illinois and again in Virginia, where he was sentenced to life in prison, confessed and pled guilty

---

[2] Blake's results and his critical appraisal of Ban's work led to one of Washington's lawyers, Peter Neufeld of the Innocence Project, calling for an audit of *all* of the Virginia lab's DNA work. Paul Ferrara, the head of the Virginia lab, dismissed Blake's work because he had been hired by the defense, defended the lab's results, and refused an independent audit. Finally, Governor Warner intervened and ordered an outside investigation.

to the rape and murder of Rebecca Williams. Later that year, Washington was formally exonerated; Governor Tim Kaine issued a full pardon, acknowledging that Washington had been wrongfully convicted of rape and murder 23 years earlier. Washington is now married and lives in Virginia Beach. He was awarded $2.25 million for his wrongful conviction from the estate of the agent who had coerced the false confession.

This case, like many of the exoneration cases, illustrates the tragic consequences of coerced confessions and the reluctance of prosecutors to fully pardon someone wrongfully convicted until another suspect has been found and convicted. It also illustrates, in the most dramatic fashion imaginable (literally, that of life and death), the difficulties and challenges involved in interpreting forensics mixtures, that is, samples with more than one contributor. The initial DNA analysis involved only one genetic marker, HLA-DQalpha, which made the results even more difficult to interpret. Even with the additional genetic markers provided by the current panel of STRs, analyzing mixed samples can still pose a challenge.

The misinterpretation of forensics mixtures lies at the heart of most miscarriages of justice and wrongful convictions attributed to DNA analysis. Many of the articles critiquing forensic genetics cite this general problem of mixtures and the Josiah Sutton case, in particular, to argue that DNA "is not infallible." "DNA's Dirty Little Secret," by Michale Bobelian in the *Washington Monthly* (2010), suggests that "a forensic tool renowned for exonerating the innocent may actually be putting them in prison." The initial analysis of the mixed forensic specimens that led to the wrongful conviction of Josiah Sutton illustrates the potential for misinterpretation. In this case, however, *misrepresentation* of the DNA results by the Houston Police Dept DNA lab and prosecutor may be a more accurate description of how the evidence was presented to the jury.

In Houston in 1998, a rape victim was assaulted by two men at gunpoint; she identified Josiah Sutton, then 16, as one of her attackers. The Houston Police Dept crime lab analyzed the crime scene samples with the then commercially available forensic DNA test kits our lab had developed: the HLA-DQalpha and Polymarker kits, illustrated in Figure 3.1 in Chapter 3, and another test that analyzed an additional genetic marker, the D1S80 locus. (The development of these and subsequent forensic tests is discussed in Chapter 5.) The results from analyzing a semen sample taken from the victim indicated a mixture of two genetic profiles; the laboratory claimed that Sutton's profile could be detected in the mixture. Sutton was convicted of rape and sentenced to 25 years based on what turned out to be an incorrect inclusion. Throughout the investigation, trial, and his subsequent incarceration, Sutton asserted his innocence and requested independent DNA testing. His requests were denied.

As it turned out, an independent investigation of the Houston Police Crime lab by two investigative reporters, David Raziq and Anna Werner, from the Houston television station KHOU, led to a reconsideration of the evidence in the Sutton case. Following up on rumors about problems with the lab's work, they used the state public record act to copy and send transcripts, laboratory records, and case reports to a group of DNA experts, including William Thompson, a Professor at University of California, Irvine, and a well-known and knowledgeable expert witness for the defense. Thompson concluded that the Sutton case report's interpretation of the HLA-DQalpha, Polymarker, and D1S80 results was fundamentally wrong. His report led to additional testing of the remaining semen sample by the private laboratory, Identigene, using the recently developed STR (short tandem repeat) technology. Their results indicated a mixture of DNA from two men, neither of whom was Josiah Sutton, a clear exclusion. Sutton was exonerated in 2004 after four and a half years in prison.

Sutton should have been excluded from the *initial* data, however, as Thompson's report points out. In addition to the misinterpretation of the DNA results, which incorrectly claimed Sutton's profile could be detected in the mixture, the lab grossly overstated the statistical significance of their analysis. The estimated frequency of Josiah Sutton's genotype for the combined HLA-DQalpha, Polymarker and D1S80 markers in the Black population is 1/694,000. This was the number presented to the jury, ignoring the complexities of interpreting mixtures and assigning random match probabilities (RMPs) to a potential contributor. Thompson's calculations indicated that "in any randomly selected pairs of black men in Texas, there is approximately 1 chance in 8 that at least one man would be "included" as a possible contributor to the vaginal sperm fraction."[3] Whether the presentation of DNA evidence by the lab in this case reflected ignorance, incompetence, or cynicism is not clear. Other work by the Houston DNA lab suggests that "all of the above" may be the correct answer and the lab was shut down by the state of Texas in 2004.

[3] *Review of DNA Evidence in State of Texas v. Josiah Sutton* (District Court of Harris County, Cause No. 800450)

Even for a competent forensic lab, analyzing forensic mixtures is challenging. The *recognition* that a forensic specimen is mixed, however is straightforward. If more than two alleles are present in a sample, then more than one individual has contributed to the specimen. Less clear, however, is the question of how many different contributors are present and how do the alleles observed in this mixture "go together "as a genotype. The interpretation depends on which assumptions are made and some of these assumptions may be questionable. If four different alleles of a genetic marker (e.g., 1,2,3,4) are detected in a blood stain, how many different individuals contributed to the mixture? At least two, but possibly, more. And how do these alleles "go together" to form the genotypes of the contributing individuals? (The simplest possible resolution would be one individual with a 1/2 genotype and one with a 3/4 genotype . . . but of course, there are many other possibilities.) One potential response for a forensic analyst confronted with a very difficult mixture is to concede that a reliable interpretation is not possible given the nature of the data and formally conclude that the DNA evidence is "inconclusive." Had I taken this approach for the vaginal swab in the Earl Washington Jr. case, he would have been executed in January of 1994.

The analysis of a simple two-person mixture when one of the contributors is known can be relatively straightforward. In the case of a mixed profile from the vaginal swab in a rape kit, the profile of the victim could be considered and the alleles of her genotype could be "subtracted" from the mixture alleles, with the remaining alleles being attributed to the unknown contributor (the rapist). The unknown contributor's profile could then be treated like that of a single source sample with the significance of the evidence calculated using the RMP statistic. For most forensic mixtures, however, this strategy is not possible. When *all* possible genotypes that are consistent with the data derived from analysis of a mixed sample are considered, the metric of significance for the evidence is expressed as the probability of inclusion (PI), the probability that the suspect's genotype is included in the mixture, rather than the RMP. If, for example, the prosecution argues that the suspect's genotype is present in the mixture, he/she *might choose to* present the population frequency of that specific profile (say, 1/10,000 for a single marker) as the RMP. However, the appropriate measure of significance should also take into account all the other possible genotypes that are also consistent with the results derived from this mixture, yielding a much less impressive metric of significance. Considering *all* possible genotypes also reduces the ability to conclude that the DNA evidence excludes the suspect. For tests with multiple genetic markers, like the STR panels, a PI for each individual marker in the profile can be calculated and then combined to generate a combined probability of inclusion (CPI).

The CPI or the likelihood ratio (LR) are clearly valuable metrics for interpreting forensic mixtures, but they require a known reference sample. Not a problem for the vast majority of cases, but what about suspect-less crimes? The current strategy for dealing with profiles from DNA mixtures, while logical, represents a paradigm shift from the initial simple concept of DNA fingerprinting and the concept of a match between a single source evidence profile and a reference profile or one generated by a database search. So, instead of interpreting the sample as a mixture consisting of the individual contributors X, Y, and Z, we can only interpret the mixture if we have the profile of a suspect corresponding to X, and calculate the probability that X is included, without considering the unknown other contributors (Y and Z).

The difficulty in the interpretation of mixtures is often cited in articles critiquing DNA forensics analysis. There is a considerable body of literature that can be characterized as "The Dark Side of DNA" journalism; many of these articles refer to two studies of mixtures carried out in 2005 and 2013, MIX05 and MIX13 (Butler et al., 2018) by the National Institute of Standards and Technology (NIST). NIST sent the same two-person and three-person contrived mixture sample to analysts from 108 labs and asked them to determine whether a suspect's DNA was present. Of the responding analysts, 70% said that the suspect might be included, 24% said the data were inconclusive, and 6% said that the suspect's DNA was not present. The exclusion was the correct answer. The range of interpretations and results revealed by this NIST survey and cited as an example of DNA fallibility in articles and courtrooms illustrates the challenges of analyzing mixtures with the then-current STR technology, particularly for mixtures with more than two contributors.

Fortunately, recent developments in statistical analysis as well as in DNA sequencing technology have made analysis of forensic mixtures more reliable. All systems of detection require the distinction of "signal" from "noise," the ability to distinguish a true detection event from background noise. For STR genotyping, an STR genetic variant (allele) is detected if the "signal," in this case a fluorescent peak in an electropherogram, is observed above some threshold value, representing system noise. (The technology of STR genotyping is discussed in more detail in Chapter 5.) For mixtures with low amounts of DNA, it is possible that a fluorescent peak corresponding to a true allele from one contributor may fall below the threshold detection level ("noise') and be missed. This phenomenon is known as "allele drop-out."

With the development of sophisticated computation methods, it is now possible to *model* these potential scenarios as well as consider *all* possible genotypes giving rise to the observed STR profiles. This strategy of "probabilistic genotyping" uses biological modeling, statistical theory, computer algorithms, and probability distributions to calculate LRs as a reporting statistic. The significance of an inclusion is expressed as the LR, given the genotype of the suspect versus that of an unknown, unrelated individual in the population. So the LR for a true contributor (inclusion) would be much greater than 1 and for a noncontributor (exclusion), much less than 1. Currently, several different probabilistic genotyping softwares are commercially available; STRMix (Taylor et al., 2013) and TrueAllele (Perlin et al., 2015) are the most widely used. The implementation of probabilistic genotyping has made the challenging problem of interpreting mixtures much less subjective and much more reliable. A recent interlaboratory study of forensic mixtures using the STRMix software (Bright et al., 2019) with 174 participants found a high degree of reproducibility, with only minor differences in the calculated LRs.

It's estimated that around 50% of all forensic samples are mixtures. The most difficult category of mixtures are those with very small amounts of DNA, sometimes known as "trace" DNA; these amounts are at the limit of detection, even with additional cycles of PCR amplification. This means that the profile of a suspect, who is truly included in the mixture, might have one or more alleles (STR peaks) that are *not* detected in the profile ("allele dropout") of the trace DNA mixture. If this were not enough of a problem, the presence of more than two contributors can further complicate the interpretation of mixtures. Different commercial softwares use different algorithms but they all must make certain assumptions about the data including interpreting "missing data" (e.g., STR peaks) that are not actually observed in the mixture but are assumed to be present and below the level of detection. For example, consider a 20-genetic-marker profile from an evidence sample that that matches the reference (suspect) profile at 19 of those markers but at one locus, the evidence profile lacks one allele (STR peak) present in the reference profile. This could be considered an exclusion but a more likely explanation is that the evidence was, in fact, contributed by the suspect and that, for some reason, perhaps the minute amount of DNA in the evidence sample, one allele in the evidence profile was missed. A probability can be assigned to this possibility. Some softwares require the assumption of a specific number of contributors to the mixture while some do not. The assumptions and statistical modeling used in STRMix and True Allele are not identical, but they typically generate very similar LRs from the same data. Analyzing the same DNA evidence with both softwares would, of course, provide additional confidence in the results. For the past two decades, both softwares have been used in many hundreds of cases, but the acceptance of the technology in the courtroom remains an ongoing issue.

In the 2016 murder case of a young boy in New York (*New York vs. Hillary*), a mixture was recovered from the victim's fingernails and a minor contributor could be detected in the STR profile but the conclusions reached by the two softwares were different. Analysis of the mixture by STRMix generated an inclusionary LR for the suspect while the analysis by TrueAllele proved inconclusive. In the "dueling softwares" admissibility hearing, Dr. Mark Perlin, the developer of True Allele, testified for the defense that STRMix analysis of this mixture was not reliable; Dr. John Buckleton, the developer of STRMix, testified for the prosecution that the results were convincing but recommended that the software be validated in-house, which was not done. In the admissibility hearing, the St. Lawrence County judge's ruling was based on neither reliability nor general acceptance by the scientific and legal community (the Frye standard) but on a procedural issue. While finding the STRMix analysis "reliable," the judge nonetheless excluded the results because the lab had not performed the necessary internal validation studies of the software.

Another significant issue in the admissibility of probabilistic genotyping results is the software code, which is something of a "black box." Some boxes, however, are blacker than others. Since 2016, STRMix has established a policy, allowing access to the source code, developmental validation records, and user's manuals to attorneys, scientists, and expert witnesses. The company that makes TrueAllele, Cybergenetics, however, maintains that its software code is a proprietary trade secret. As of 2019, TrueAllele has withstood 27 admissibility challenges and remains widely used but, recently, judges in two cases in Pennsylvania and Virginia (2021) have sided, for the first time, with defense requests to see the source code. Issues of admissibility notwithstanding, the implementation of probabilistic genotyping has made it possible to resolve mixed STR profiles previously thought too complicated to interpret, a critical contribution to forensic genetics and to the criminal justice system.

Recent technological developments in DNA sequencing, known as next generation sequencing (NGS) or, in forensic circles, as massively parallel sequencing (MPS), have also simplified the analysis of mixtures. (I discuss this technology, as well as other recent developments in more detail in Chapter 5, "The Weight of Evidence: Statistics and the Evolution of Forensic Genotyping.") One critical element of NGS, which was first introduced in 2005, is "clonal" sequencing. The other critical feature of NGS is that it is "massively parallel," as the platform developers refer to the millions of DNA sequencing reactions, all initiated with a single DNA molecule (i.e., clonal), that are performed in parallel. The previous DNA sequencing system, known as Sanger sequencing after Fred Sanger, the Nobel laureate who developed the technology, would generate a *composite* sequence of the different DNA sequences in the mixture. Unlike Sanger sequencing, each sequencing reaction in NGS starts with a single DNA molecule, so each individual component of a mixture is sequenced *separately* in parallel reactions. With Sanger sequencing of an 80%/20% mixture of two individuals, the results might show at a given position in the DNA sequence a mixed base, e.g., an A and a T, and at a nearby position another mixed base, e.g., C and T. With NGS analysis of this same sample, around 80% of the individual clonal sequences that were generated would have an A and C while around 20% of the sequences would have a T and T at those positions

In addition to the very valuable clonal sequencing property of NGS, focusing on genetic markers like polymorphisms in the mitochondrial DNA (mtDNA) or Y-chromosome, known as "haploid lineage" markers because each individual has only one copy (haploid) rather than the two copies (diploid) of an autosomal gene, can help resolve a mixture. So, for example, a three-person mixture should have three different mtDNA sequences or three different Y-chromosome sequences, while it would have from three to a maximum of six different chromosomal gene variant sequences (alleles) for each chromosomal locus. The very considerable challenge of how to assemble these variant sequences into three genotypes, one for each contributor, is greatly simplified if only haploid lineage markers are used in the analysis. Mitochondrial DNA has another uniquely useful property for forensic analysis. Mitochondrial DNA is contained within the mitochondrion, the cellular organelle that is responsible for energy generation within the cell; each cell contains thousands of copies of the mitochondrial genome as opposed to the two copies per cell of a chromosomal gene. As a result, mtDNA is often the *only* genetic marker used to analyze samples with minute amounts of

DNA. This property is why mtDNA was the first genetic marker successfully PCR amplified and analyzed from the DNA of ancient remains, as discussed in Chapter 10. It is also why mtDNA is valuable in the search for missing persons, because the small amount of DNA that can be recovered from the victim's bones and used for identification is usually degraded. When Mark Stoneking, the evolutionary biologist now at the Max Planck Institute in Leipzig, was a visiting scientist in my lab at Cetus, we carried out the first identification of a missing person using a new PCR-based mtDNA test to analyze the skull bone fragments of a missing child in 1991, discussed in more detail in Chapter 8.[4]

One might ask, if mtDNA is so useful for analyzing forensic mixtures, why isn't it used all the time? One reason might be that sequencing mtDNA is more complicated and more expensive than the current standard STR technology. The main reason, I suspect, is that a *match* or *inclusion* with mtDNA is much less discriminating than a match with the current set of chromosomal genetic markers. The polymorphic sequences in the mtDNA genome are physically linked on the same molecule and, therefore, not statistically independent, as are the chromosomal genetic markers used in individual identification. Consequently, the "product rule," that allows the calculation of very low matching probabilities by multiplying the probability of matching at one locus by the probabilities at all the other loci in the test is not applicable for mtDNA, nor for Y-chromosome genetic markers. The numerical significance of a match for mitochondrial DNA (or Y-chromosome markers) is determined, in part, by the size of the population database. So, if the mitochondrial sequence of a hair found at the crime scene matched the sequence of a suspect and, in a population database of 10,000 mitochondrial sequences, this particular sequence was found twice, then the estimated RMP would be around 1/5,000, a far cry from the < 1/100 million RMP that can, in many cases, be generated using the standard STR technology and that juries have come to expect.[5] The other limitation of mtDNA as a genetic marker of individual identification is that, since the mitochondria are present in the egg and, therefore, inherited only through the maternal line, all maternal relatives share the same mitochondrial DNA sequences. Therefore, if the

---

[4] Stoneking et al. (1991).

[5] Probability is not always equal to frequency, as my friend, Charles Brenner, a forensic mathematician, likes to point out. He discusses how to estimate the RMP if the mitochondrial DNA sequence (or the Y-chromosome marker) found in the evidence sample and in the suspect is *not* present in the database. In such cases, the RMP can be significantly lower than 1/n, where n = the size of the database.

mitochondrial DNA profile of a forensic evidence sample matched the suspect, it would also match his sibs, his mother, and all other maternal relatives (uncles, aunts, grandmothers). This feature, which limits the discrimination power of mitochondrial DNA analysis, can be useful in the identification of missing persons or human remains, as the reference sample can be provided by any maternal relative.

These limitations notwithstanding, mtDNA is enormously valuable for the analysis of forensic samples that have very little DNA and for mixtures, and particularly powerful when the analysis uses NGS. Many forensic specimens, like cartridge casings, can be analyzed thus far only with mtDNA.[6] It allows determining the number and the relative proportions of contributors to a mixture and to identifying the contributors with many fewer assumptions than are required with interpretations of chromosomal markers. Our lab is currently working on an NGS-based strategy to analyze the same forensic sample with both the mtDNA sequence and chromosomal genetic markers (short tandem repeats or STRs and single nucleotide polymorphisms or SNPs).[7] Applying this strategy to forensic mixtures can achieve the resolution provided by mtDNA as well as the discrimination power derived from the hundreds of chromosomal markers.

[6] Wisner et al. (2021).
[7] Bali et al. (2021). California Association of Criminalists

5

The Weight of Evidence

Statistics and the Evolution of Forensic Genotyping

The prosecutor tells the jury that the genetic profile of the crime scene evidence matches that of the suspect. This statement seems straightforward, . . . but I would guess that relatively few jurors (or readers) are familiar with how the statistical metrics used to interpret the DNA evidence are actually calculated and what assumptions are used in those calculations. Genetic testing of crime scene evidence predates DNA technology, but the logic of interpreting matches remains the same.

The Interpretation of Matches

Before DNA analysis entered the courtroom, comparing patterns obtained from evidence samples to a reference sample from a potential suspect was performed by blood group typing or analysis of the physical properties of proteins found in the sample. The critical element in these comparisons was the genetically determined variation in the human population of these patterns, either blood group types (A, B, AB, O) or the size and electric charge of specific proteins. The basic logic of this approach is simple: Does the evidence pattern match the reference pattern? If not, the evidence could not have come from the suspect, a result known as an "exclusion." If the patterns are indistinguishable (i.e., an "inclusion"), the evidence specimen *could* have come from the suspect, but it could also have come from someone else with the same blood type. An exclusionary result requires no statistics, but the meaning and weight of an inclusion (match) depend on the probability of someone other than the suspect also matching the genetic profile of the evidence specimen. The probability of such a coincidental match is known as the random match probability (RMP), the probability that an individual randomly chosen from the population would happen to have the same genetic profile. This probability is related to the frequency of a specific genetic profile in the population, so interpreting a match requires a population database. Since some genotypes are more common in the population than others, a match with a rare genotype would be more significant than a match with a more common genotype.

While the RMP is a statistic that is often presented to juries, an alternative metric is the likelihood ratio (LR), based on the DNA evidence. The LR is the ratio of the probabilities of the evidence, given the prosecution's hypothesis (the evidence came from the suspect) and the defense's hypothesis (the evidence came from someone else). In the analysis of a *single source* sample, the LR is, typically, the reciprocal of the RMP because the probability of the evidence given the prosecution's hypothesis is 1 and the probability of the evidence coming from a coincidental match (the defense's hypothesis) is the RMP.

The RMP or the LR is calculated for a *specific* genotype, the one found in both the evidence and reference samples and these are the metrics presented to the jury. A metric that the jury will never see but one that allows comparing the potential for a particular genetic test to distinguish individuals is known as the power of discrimination (Pd). This is defined as the probability that two individuals randomly chosen from a relevant population would prove to be *different* based on the test results, considering *all possible genotypes* at a genetic locus. For the HLA-DQalpha test, the first PCR-based marker used in the 1986 *Pennsylvania vs. Pestinikas* case, discussed in Chapter 1, and the first commercial PCR-based forensics genotyping kit, launched in 1991, the Pd was relatively modest. Using the currently available standard PCR commercial kits for 23 different genetic markers, the Pd is vastly more impressive. As we all know from newspaper accounts, the RMP for a particular multilocus genetic profile might be on the order of 1/10 billion. Forensics statistics analysis is inextricably linked to the evolution of forensic genotyping technology, and so I have placed my discussion of these two issues in the same chapter.

Obviously, the significance of the RMP for a guilty versus innocent verdict is dependent on other non-DNA related evidence. A RMP of 1/1,000 or even 1/100 combined with other reasonably compelling evidence would be very persuasive, while the same RMP in the absence of other evidence would not. The RMP "threshold" for a conviction without any other evidence would have to be very low indeed. As the technology has evolved and additional genetic markers have been developed, the bar for a "meaningful" or persuasive match has risen. Prior to the introduction of DNA evidence, matches based on genetic systems, like blood groups or the electrophoretic mobility of certain proteins, were considered to be highly significant. The metric for the potential to distinguish two different samples, the discrimination power, for these non-DNA genetic systems *combined* was less than 1/100 with correspondingly modest RMPs. Now, juries have come to expect RMPs that are many orders of magnitude lower.

During the late 1980s, when PCR genetic typing was being introduced into forensics, a common view was that it was an impressively powerful tool, allowing the analysis of difficult forensic specimens (with little and/or degraded DNA) but that it was much less discriminating (i.e., had a lower Pd,) and was therefore less informative for inclusion than RFLP analysis of the variable number tandem repeat (VNTR) loci. This perspective was based on the availability at the time of only one PCR genetic marker, the HLA-DQalpha locus, compared to multiple genetic markers, the VNTR loci, for RFLP analysis.

The initial DNA fingerprinting analysis in 1985, pioneered by Alec Jeffreys, used a radioactive DNA probe that recognized many different parts (the VNTR loci) of the genome, resulting in a very complex pattern; consequently, matches were difficult to interpret statistically. Subsequent development of the technology yielded multiple "single-locus" probes so that a much simpler pattern that could be interpreted statistically could be obtained for each individual probe and each individual VNTR locus. Although defining a "match/inclusion" was still somewhat complicated using the multiple single locus probe RFLP system, the significance (RMP) of a match that was estimated by multiplying the probabilities calculated for each individual locus was very impressive, on the order of 1/millions to 1/billions. In the United Kingdom, this strategy was implemented by the Forensic Science Service and in the United States, this RFLP technology was developed and applied to casework by two companies, Lifecodes and Cellmark.

PCR genetic typing was much more robust, faster and easier to perform, less expensive, and required much less DNA. Unlike RFLP, it did not require radioactive labels. It could also analyze many forensics samples, such as those with degraded DNA, as in the *Pennsylvania vs. Pestinikas* and *Dotson* cases, that RFLP could not. The solution to this dilemma was clear: develop more PCR-based genetic markers and have the best of both worlds.

So we started. The second PCR-based forensics test was based on five different genes which had sequence variants that could be distinguished by DNA probes. The Polymarker test launched in 1993 was designed to be used in conjunction with the HLA forensics kit (Figure 3.1 in Chapter 3). All of the six genetic loci could be amplified by PCR in a single tube.

As we've discussed, the informativeness of a given test for individual identification, the Pd, is the probability that the genetic profiles of two randomly chosen individuals will be different, is calculated for the distribution of all possible genotypes in the relevant population. These six genes were statistically independent of one another, so the probabilities of a match at each of the individual loci could be multiplied. The combination of the Polymarker and HLA-DQalpha tests greatly increased the Pd for PCR-based tests, but it was still less than that of multigene RFLP analysis.

The Marriage of PCR and VNTRs

The HLA-DQalpha and Polymarker tests were based on detecting variation in the *sequence* of the amplified DNA, but PCR could also be used to distinguish *length* variation, that is, variation in the number of tandem repeat copies for a given genetic segment. If the PCR primers were designed to flank the tandem repeat copies, the length of the PCR product generated by amplification would vary according to the number of copies. In keeping with the tradition of acronyms in molecular biology, these first PCR-based tests for length variation were called "AmpFLPs." The first commercial PCR test for a length polymorphism was our D1S80 test kit, which analyzed variation in the number of copies of a tandem repeat region of a specific 16bp sequence. Although it was widely adopted, there were two limitations associated with this test. Based on the detection of the amplified DNA fragments by gel electrophoresis, the D1S80 test was relatively labor intensive and required staining the gel to visualize the pattern of the DNA fragments. In addition, there was a concern that D1S80 variants that had fewer copies of the repeat and, therefore, generated shorter PCR amplification products, would amplify more efficiently than variants that had many copies of the repeat, leading to difficulties in interpretation.

Both of these issues were addressed in the development and application of the short tandem repeat (STR) markers, commercialized by Applied BioSystems (ABI, now Thermofisher) and Promega. The primers for amplifying these four base pair repeat regions scattered around the genome could be labeled with one of four fluorescent dyes and the size-based separation of the variable length PCR products was automated by capillary electrophoresis instruments made by ABI. However, the automated analysis of STRs was not without technical issues. So-called stutter bands, artifactual minor PCR products, could complicate interpretation of the results. Nonetheless, automated genotyping analysis using STR panels became and continues to be the standard method for forensic DNA analysis. The first commercial panel in the United States analyzed 13 different STR markers amplified together in a single PCR test. This panel was used for many years as a standard genotyping method for forensic specimens and for the criminal and missing person databases, supported by the FBI.[1] In Europe, another set of STR markers was used. Currently, the commercial kits include 24 different STR markers that include the initial core sets used in the United States and Europe.

The STR markers, like the HLA genes, have many different alleles but the mechanisms underlying the length (STR) and the sequence (HLA) polymorphism are very different. During DNA replication, the number of copies of short (four nucleotides) tandem repeats can increase or decrease. Thus, STRs are highly polymorphic because they are highly *mutable*. The sequence polymorphism in the HLA genes, however, is relatively stable; the allelic diversity is the result of balancing natural selection, which maintains many different alleles in the population, rather than the result of a high mutation rate. The different HLA sequences are functional, allowing individuals and populations with more HLA diversity to mount a more effective immune response to more different kinds of pathogens.

The STR markers, analyzed with capillary electrophoresis and genotype-calling software, became the dominant forensic genetic markers, gradually replacing the sequence-based genetic markers we had developed. The initial panel of 13 STR markers was much more informative (i.e., discriminating) than our original sequence-based tests (HLA-DQalpha and Polymarker). Also, since the HLA sequence polymorphisms were functional and contained potentially relevant medical information, raising concerns about privacy, the STR markers were promoted as being in "junk" DNA regions that did not code for proteins. Presumably, therefore, they conveyed no information about physical traits or medical status. The same arguments about potential medical information have been raised by some critics about

---

[1] The Combined DNA Index System (CODIS) is the term used to describe the criminal justice DNA databases as well as the software for running these STR databases. The collection of DNA profiles contributed by federal, state, and local participating forensic laboratories is known as the NDIS (National DNA Index System).

the use of the parts of the mitochondrial DNA genome that encode proteins for individual identification. Since, in the pre-DNA era, blood groups and proteins were the forensic genetic markers in routine use, this critique of protein coding DNA polymorphisms seemed less than compelling to me, but this argument served to protect the nascent field of DNA genetic typing from many of its critics. *Nothing to see here, folks! Just some junk DNA. Let's just move on.*

I was also a bit skeptical about the argument that noncoding genetic markers were completely free of medical information and that therefore we should dispense with all other genetic markers, like HLA, however useful. Although the STR markers do not directly encode any relevant medical information, they *could be* correlated with nearby functional genes that did.[2] Moreover, some noncoding DNA has "regulatory" functions, that is, it influences gene expression and protein synthesis. A recent analysis of 150,000 human genomes in the UK Biobank by Kari Stefansson and colleagues from the Icelandic genetic company deCODE demonstrates that the vast majority (89%) of disease-causing variants occur in these noncoding regulatory elements of the genome (Halldorsson et al., 2022).

So, although the distinction between coding ("bad") and noncoding ("good") DNA seemed oversimplified, playing the "junk" DNA card was, at the time, a convenient and useful political argument for the acceptance of forensic DNA technology in its contentious infancy. For the forensic community, the release of these standardized commercial STR kits and semiautomated electrophoresis instruments transformed genetic typing for both evidence samples and databases.[3] As the number of STR markers in these kits increased from the original 13, the power of discrimination skyrocketed.

---

[2] A recent report (Banuelos et al., 2022) demonstrated that some of the forensic STR markers were, in fact, associated with the expression of neighboring genes and with potential medical information. This observation was presented as a legal "bombshell" but is the result I would expect from any study focused on investigating these potential associations. We can now no longer rely on the "convenient fiction" that the STR markers convey *no* medical information and need to preserve and protect privacy, to the extent possible, by other means.

[3] One of the reasons that the STR markers prevailed from a commercial perspective was that Applied BioSystems, the main company that manufactured and distributed the new STR kits, also manufactured and distributed the capillary electrophoresis instruments used for DNA sequencing that could now be used for genotyping samples for length polymorphisms. Applied BioSystems also distributed the sequence-based forensic tests we had developed and manufactured, but those tests did not require the use of their instruments. The profit motive is uniquely persuasive; with this incentive, Applied Biosystems' commercial focus turned to the STR systems and away from the probe-based kits for analysis of sequence polymorphism.

In our lab, we continued to develop the PCR-based tests for all the different HLA genes for use in matching donor and recipients for bone marrow and solid organ transplantation and to carry out basic research in genetic susceptibility to autoimmune diseases, such as type 1 diabetes and multiple sclerosis. I remained fascinated by the role of HLA polymorphism in the immune response and analyzing these genes continued to be the focus of my lab's research. But, for the forensic community, the momentum for genetic typing had clearly shifted to the analysis of length polymorphism and the commercial STR kits, and they have remained the standard for forensic genetic typing.

As we've seen, the more genetic markers available in a given genetic test, the more discriminating the test will be. The calculation of the RMP for a given genetic profile (e.g., a 16-locus match) is based on the observed frequencies of the 16 individual locus genotypes in the "relevant" population. These population databases used for determining frequencies are fundamentally different from the criminal databases used in "cold hit" searching (see Chapter 6 on databases). They are presumed to represent a random sample of individuals from a given ethnic population. So a particular genetic profile might have a frequency a in a European American (Caucasian) population, b in an African American population, c in a Hispanic population, d in an Asian population, and e in a Native American population. Clearly, these are very broad categories and the definition of the "relevant" population will depend on the circumstances and location of the crime. So there can be some variation in the calculated RMPs, based on the database used for the calculation; that's why they are called *estimates*. These issues are discussed in more detail in Chapter 6.

These estimates of RMPs are based on specific assumptions, one of which is statistical independence of the different genetic markers used in calculating the frequency of a particular multilocus genetic profile in the relevant population.[4] The databases used in calculating these multilocus genetic profile population frequencies are assumed to be a random representation of the

---

[4] The classic example of flipping a coin multiple times illustrates the concept of statistical independence. If the probability of tossing a head is ½ and the probability of a head on the subsequent flip is independent of the first, then the probability of tossing 4 heads in a row is estimated as ½ × ½ × ½ × ½ or 1/16, assuming independence of all flips. Similarly, using the so-called product rule, the frequency of a multilocus (e.g., 6 different markers) profile is calculated by multiplying the frequency of the profile at locus 1 (e.g., 1/10) times the frequency of the locus 2 profile (e.g., 1/10) times the frequency of the profiles at the other 4 loci (all 1/10), generating a frequency estimate of 1/10 to the 6th power, or 1 million.

population. It is the assumption of statistical independence and the use of the product rule that results in the calculations and presentations to the jury of RMPs of <1/billion that have been generated from population databases of only thousands of individuals.

Some of the earliest critiques of DNA forensic typing challenged the assumption of statistical independence among genetic markers and the use of the product rule in generating estimates of the RMP and/or the LR. The statistical independence of genetic markers, a requirement for using the product rule to calculate RMPs, is known as "linkage equilibrium"; demonstrating the statistical independence in population validation studies is a prerequisite for any set of genetic markers to be used and accepted in forensics analyses. The traits blue eyes and blond hair represent a familiar example of a deviation from linkage equilibrium within a population; suppose the frequency of blue eyes is 1/10 and the frequency of blond hair is also 1/10. Assuming independence, one would estimate that the frequency of these two traits together in the population would be 1/100. But of course, the observed frequency of these two traits together is just slightly less than 1/10, given the low frequency of dark haired individuals with blue eyes and blond haired individuals with dark eyes.

Deviation from linkage equilibrium is one of the reasons we didn't follow up our initial forensic genotyping test for the HLA-DQA1 locus with tests for the other even more polymorphic HLA loci to increase the discrimination power of our forensic tests. Genotyping *all* the highly polymorphic HLA loci[5] was extremely valuable for our immunogenetics studies but was not well suited to forensic individual identification because the allele frequencies for these tightly linked loci were not statistically independent and, consequently, the product rule could not be used in calculating RMPs.

The estimation of RMPs is based on the frequency of a particular multilocus *genotype* but the forensic population databases used in the calculation contain information on *allele* frequencies. Estimating genotype frequencies from allele frequencies is based on another assumption that is less intuitive than the statistical independence of genetic markers (linkage equilibrium) but that may, nonetheless, be familiar to readers with a vivid recall of high school biology. Hardy-Weinberg (HW) equilibrium, named after a mathematician and a biologist who, independently, had the same fundamental

---

[5] The most widely analyzed HLA loci are HLA-A, -B, -C, -DRB1, -DQB1, and -DQA1, the locus used in the first forensic genotyping test, the HLA-DQalpha forensic kit. These loci are the most polymorphic in the human genome; some of these loci have thousands of alleles.

insight, describes the relationship of genotype and allele frequencies in randomly mating populations.[6] The reason that HW equilibrium is so important for forensics RMP estimates is that it allows the calculations of genotype frequencies from observed allele frequencies in the population databases. Frequency estimates for alleles will be less subject to sampling error and, therefore, much more accurate than for genotypes simply because there are so many fewer alleles. Consider a genetic marker with 10 different alleles; there are 45 different possible genotypes, that is, specific combinations of these 10 alleles. In a database of, say, 10,000 individuals, the observed frequency data for 10 alleles will be much more reliable because it represents a more accurate sampling of the population than would the observed genotype data.

So, in the calculation of the RMP, the genotype frequency for a given genetic marker (say, locus 1) is calculated, using HW equilibrium assumptions, from the observed allele frequencies for locus 1 as well as for all the other loci. Then, using the product rule, the frequency of a specific multilocus genetic profile is calculated. Demonstrating that HW equilibrium holds for a given genotyping technology applied to a randomly mating population is considered a critical validation for that technology. HW equilibrium is evaluated by comparing the *observed genotype* frequencies determined experimentally with the *expected* genotype frequencies calculated, assuming HW proportions, from the *observed allele* frequencies. For a given population, a significant deviation from HW equilibrium using a validated genotyping technology can be interpreted to mean that the population in question may contain discrete subpopulations and that, therefore, the frequency estimates, based on the assumption of HW proportions, may be inaccurate. In the early days of forensic DNA applications, several critics noted that, in some RFLP (VNTR) population databases, there was an excess of observed homozygotes compared to the number expected under the HW equilibrium assumption of a randomly mating population. This excess could, in theory, be explained by population substructure. A heated and vigorous debate about this issue, namely, the potential of population substructure to give rise to inaccurate frequency estimates, led to a National Research Council report in 1992, recommending what, in retrospect, were overly

---

[6] For a simple two-allele (e.g., A and B) genetic marker, the frequency of the three possible genotypes AA, AB, and BB can be estimated from the observed allele frequency for A (p) and for B (q) as AA = p2, AB = 2pq, and BB = q2.

conservative statistical guidelines but ones that allowed forensics DNA analysis to continue (see Chapter 2 and the DNA Wars).[7] (Conservative, in this context, means using RMP estimates that are higher due to a "correction factor" for *potential* population substructure.) In 1996, another National Research Council Report reexamined the issue of population substructure and RMP estimates, concluding, based on extensive population data, that the conservative guidelines recommended in the previous report were unnecessary. Currently, HW equilibrium analyses are used primarily to examine the reliability of a particular genotyping technology.

So, the RMP that is calculated from these population databases, like all *estimates*, involves some simplifying assumptions and, thus, some uncertainty. One presumption is that, provided the estimated RPM is sufficiently low, say, <1/10 billion, a little uncertainty in the RPM estimate does not pose a major problem, and a number like this could be taken, and often is, as evidence of unique identification.[8] This presumption, while not unreasonable for most samples and for most cases, can be, however, more complicated and potentially problematic for partial genetic profiles due to degraded DNA or specimens with multiple contributors (mixtures).

There are many legitimate and scientifically interesting issues to consider in how the statistical probability of a random match is estimated and how this number (or numbers) should be presented to the jury. In the early days of DNA analysis, some judges and courts decided, based on their discomfort with how RFLP matches were defined and how RMPs were estimated, that exclusionary results but not inclusionary results were admissible and could be presented to the jury, as we saw in Chapter 2 and the 1989 *People vs. Castro* case.

So . . . *exclusions were included while inclusions were excluded.*

Consistent with this view, defense attorneys or sophisticated expert witnesses often made plausible arguments that, given the assumptions made in calculating genotype frequencies, the RMPs for a given 16-marker inclusionary profile have an element of uncertainty. They then argued, less plausibly, that, since there is some uncertainty in the statistical estimate of the

---

[7] Ironically, a critical study demonstrated, for some RFLP markers, that the deviation from HW equilibrium, which had occasioned all the population genetics brouhaha and the National Research Council Report (1992) guidelines, was due to a problem with the RFLP technology causing heterozygous samples to be misclassified as homozygotes and not to population substructure.

[8] Monozygotic twins cannot be distinguished by the standard STR panels; next generation sequencing of the whole genome, however, can distinguish them.

RMP, it is therefore misleading and inappropriate to present this number (*or any number*) to the jury. In my view, it would serve the justice system better to acknowledge that these numbers are *estimates* and allow a discussion by statistical experts to be presented to the jury.

If the prosecution presents a RMP of, say, 1 in a million, an estimate based on calculations that make certain assumptions, the defense might argue that, making different assumptions, the estimate might be closer to 1 in 10,000. Although 1 in a million does not constitute an absolute individual identification—there are many hundreds of people on the planet with the same genetic profile—it is, nonetheless, an impressive number and, based on an argument by the defense of innocence and a coincidental match, the allegedly innocent suspect would have to be remarkably unlucky to share a very rare genetic profile with the actual perpetrator. It is also, presumably, very unlikely that any of those individuals with this very rare profile would also have opportunity and motive. On the other hand, 1 in 10,000 is a different story. The defense theory might be that some other individual with this genetic profile was the source of the evidence and that the suspect was innocent. One possible response would be to say, OK, we accept that there are questions about the assumptions made in this estimate and that this inclusionary (or failure to exclude) result may be due to chance alone. . . . In that case, provided that there is sufficient DNA from the forensic sample left, let's genotype an additional 2 or 3 markers and *surely*, your innocent client, who unfortunately just happened to match the evidence at these 16 loci, will be excluded by additional analysis. This approach could separate those arguments that are based on a belief in the client's innocence from those that represent an attempt to prevent the RMP from being presented to the jury.

Beyond STRs: Recent Developments in Forensic DNA Technology

As we've seen, STR genotyping using capillary electrophoresis to separate the amplified DNA fragments continues to be the standard technology used in forensic cases. However, over the past several years, different kinds of genetic markers as well as novel technologies to genotype these markers have been applied to difficult crime scene evidence samples, like mixtures or highly degraded DNA.

The single nucleotide polymorphisms (SNPs) scattered throughout the genome have been used for many years as genetic markers in medical and population genetics research; they are particularly valuable in forensic analysis when the DNA is too degraded for STR analysis. Several different DNA technologies have been used to determine SNP genotypes but, currently, the most common is the probe-based microarray, a technology widely used by the very large medical genetics projects known as genome wide association studies and the commercial DNA testing companies, like 23andMe and Ancestry.com. This genotyping method, like the DQalpha and Polymarker tests, is based on the principle of DNA hybridization and the binding of specific DNA probes, short synthetic pieces of DNA, to determine the sequence of a specific DNA region.[9] Unlike the early forensic kits, the DQalpha and Polymarker tests, which had only 21 probes, the current microarrays have millions of probes on a tiny chip for genotyping hundreds of thousands of SNPs. Microarray SNP genotyping is fast, relatively inexpensive, and capable of generating huge amounts of data. Unlike the STR markers, which have been chosen to be statistically independent of one another, the SNP markers in the microarrays are often closely linked on the same chromosome, and therefore the product rule cannot be used. These SNP markers can, however, be valuable in estimating how much DNA is shared between distant relatives. Estimating the length of shared genomic regions and the reconstruction of family trees is a critical feature of investigative genetic genealogy, the strategy of searching databases to find distant relatives of the individual who left the crime scene evidence. In the criminal justice system, thus far, SNP genotyping with microarrays has been mainly used in cold cases searching genealogy databases, a topic we discuss in more detail in Chapter 6. Some forensic companies offer panels of hundreds or thousands of SNP markers, explicitly chosen to be statistically independent so that the product rule *could* be used to estimate RMP or LR. One technology for genotyping these SNP panels is next generation sequencing.

---

[9] Single-stranded DNA can bind or "hybridize" to a complementary strand, based on the base pairing of A and T and of G and C. Under the appropriate conditions, a short DNA probe will bind to its complementary sequence only if it is completely matched. Thus, the binding of a probe can be used to determine the sequence of a specific DNA segment. So, for an A/T SNP in a specific genomic region, the DNA amplified by PCR from that region in the sample is tested with probes specific for the A allele and for the T allele. Based on the observed probe-binding pattern, we can determine whether the SNP genotype is A/A, A/T, or T/T.

Next Generation Sequencing

By far, the most transformative technology development in molecular genetics in recent years has been the implementation of what is known as next generation sequencing (NGS), or in the forensic community, massively parallel sequencing (MPS). Although there are many different instruments and commercial platforms, all NGS systems have two critical features in common: (1) they can carry out millions of sequencing reactions simultaneously (hence, "massively parallel") on a chip or a flow cell and (2) each sequencing reaction is *clonal*, that is, initiated with a single DNA molecule.[10]

The massively parallel feature of all NGS systems means that huge amounts of sequence data can be generated quickly and relatively cheaply, truly revolutionizing molecular genetics research. Entire genomes can now be sequenced in days. All of the transformative sequencing of archaic and ancient human genomes, discussed in Chapter 10 ("DNA of the Dead: Sequencing Archaic Species and Ancient Remains") was made possible by this massively parallel technology. The other critical feature of all NGS systems is clonal sequencing, the analysis of a single DNA molecule or a clonal population derived from that molecule. This feature makes NGS uniquely well suited to the analysis of mixtures, one of the most challenging kinds of crime scene evidence, because the individual components of a mixture can be sequenced separately.

Consider a blood stain with three different contributors. If we were to sequence mitochondrial DNA from this mixture using the pre-NGS and nonclonal sequencing technology (i.e., the standard Sanger sequencing), we would generate a composite sequence of three different individuals. As we've seen in the previous chapter, "The Earl Washington Jr. Case and the Problem of Mixtures," mitochondrial DNA is a *haploid* genetic marker; every individual has only *one* mitochondrial DNA sequence.[11] If, however, this

---

[10] Some NGS systems start with a single DNA molecule, create millions of copies with an amplification system, and then sequence this clonal population (e.g., Illumina), while other systems sequence the single DNA molecule directly (e.g., Oxford Nanopore). Some systems, like Illumina, provide only short sequences, less than 400 bases, while others, like Oxford Nanopore, are capable of generating much longer sequences (many thousands of bases). Many years ago, when I first heard presentations about the early attempts to develop these NGS systems, I was impressed with the ingenuity and ambition of these fledgling projects but deeply skeptical that they could ever work. For the past 15 years, all of my research in both medical and forensic genetics has relied on these NGS systems. Skepticism may be a critical element of scientific analysis but, fortunately, creativity, innovation, and persistence are also hallmarks of the scientific enterprise.

[11] The issue of mitochondrial DNA heteroplasmy is discussed in Chapters 4 and 8.

mixture were analyzed with an NGS system, generating, say, 100,000 independent individual sequence reads, we might find that 60% of the sequence
reads correspond to individual 1, 30% to individual 2, and 10% to individual
3. So this NGS analysis not only identifies three different individuals in the
mixture but can estimate their proportions as well.[12]

In forensic applications, NGS analysis can be focused on specific polymorphic markers, such as a panel of hundreds of SNPs, a panel for STRs,
and the mitochondrial genome. The Pd for these combined genetic markers
would vastly exceed that of the standard STR markers analyzed by capillary
electrophoresis. There are two alternative strategies for such "targeted" NGS.
One can selectively amplify specific parts of the genome using PCR or one
can "capture" specific genomic regions by hybridizing the genomic DNA to
probes that bind to and "capture" the targeted region. Each of these strategies
has its strengths and weaknesses, depending on the specific application.

All NGS systems involve the creation of "libraries" by adding short DNA
sequences known as adapters to the targeted DNA; these adapters are required to guide the targeted DNA to the individual sites on a chip or flow cell
where the sequencing reactions take place.[13] For PCR-targeted libraries, one
can design PCR primers that flank a panel of SNP or STR markers and, for
analysis of mitochondrial DNA polymorphism, primers that allow the amplification of the whole mitochondrial genome. Forensics genetics companies,
such as Verogen, spun out of the NGS commercial leader, Illumina, in 2017,
offer PCR-based kits for NGS analysis of SNPs for individual identification as
well as for biogeographic ancestry. The use of ancestry informative markers
(AIMS) for investigative leads is discussed later in this chapter as well as in
Chapter 6 (on databases).

For capture targeted libraries, one adds the adapters to *all* the genomic
DNA to make a whole genome library and then designs specific probes to
capture the targeted polymorphic regions. My colleague Sandy Calloway and
I chose the capture strategy for forensic samples primarily because the DNA
from many crime scene evidence specimens is highly degraded, that is, is

[12] NGS can also provide critical information in the analysis of a single individual for the diploid
chromosomal markers. We can think of a single individual as a "mixture" of two alleles, the maternal and paternal. Consider a gene with two linked SNPs where both the crime scene evidence and
suspect are heterozygous A/G for the first SNP and C/T for the second. By Sanger sequencing or by
microarray genotyping, the composite sequences would be indistinguishable. But by NGS analysis,
which sequences each allele independently, the suspect might turn out to be A-C on one allele and
G-T on the other while the evidence sequence might be A-T on one allele and G-C on the other allele,
resulting in an exclusion.
[13] These adapter sequences can also serve as PCR priming sites to amplify the whole library.

present only in short DNA fragments. Very short DNA fragments are unlikely to retain both of the two intact PCR priming sites necessary for PCR amplification. Also, we could make a whole genome library from the forensic sample and then sequentially use our SNP and STR probe panels and our mitochondrial genome probe panel to capture and analyze all these genetic markers from the same sample.[14] This strategy of starting with a library of all the DNA in the sample also allows for sequencing the whole human genome, without capturing specifically targeted regions.[15] One might ask: Why sequence the whole genome when less than 1% is polymorphic and, therefore, more than 99% of the generated sequence will be un-informative for individual identification? As the cost of whole genome sequencing (WGS) decreases, this "inefficiency" becomes less of a problem and, although targeted NGS will be used much more frequently, WGS for individual identification may be applied in special cases.

As it turned out, the first admissibility hearing for NGS involved the use of WGS in the analysis of monozygotic twins (*Commonwealth of Massachusetts vs. Dwayne McNair*, 2017). In 2004, two women were assaulted and raped by two men in Suffolk County, Massachusetts. The STR profile of the crime scene evidence, a semen sample, matched that of Dwayne McNair, but it also matched that of his brother, a monozygotic twin, complicating the use of DNA evidence in the prosecution's argument. The second man suspected of the assault and rape, Anwar Thomas, worked out a plea deal and identified Dwayne McNair as the perpetrator. In 2011, Dwayne McNair was charged with aggravated rape and armed robbery, but the district attorney's office still hoped to be able to use the DNA evidence and establish definitively which twin was the source of the semen sample. They contacted a European genetics company, Eurofins Scientific, which had just published the results of their twin study (Weber-Lehmann et al., 2014), in which, using WGS, they were able to identify mutations in one twin that were passed down to his child but were not present in his "identical" twin brother. When Eurofins sequenced the genomes of the McNair twins as well as the semen sample, they found nine sequence differences between the two twins and that the whole genome sequence of the evidence was closer to Dwayne McNair. However, during the admissibility hearing, the presiding judge granted McNair's motion to

[14] Bali et al., 2021, California Association of Criminalists conference.
[15] The library from a sample may contain non-human DNA as well, such as bacterial DNA. Aligning all the NGS sequence reads in the library to the reference human genome sequence creates a dataset with only the human sequences.

exclude the DNA evidence, finding that the Eurofins genome sequencing by NGS did not meet the *Daubert* standard of admissibility. Eurofin's initial twin study, in which they found that mutations between twins were passed down to their children, involved only one pair of twins. In addition, the judge noted that NGS had not been regularly employed in criminal cases and ruled that more tests simulating crime scene conditions were needed to satisfy the *Daubert* standard. The judge's concern about the Eurofins sample size ($n = 1$) has been addressed recently by a massive study carried out by the Icelandic genetics company deCODE. The deCODE team sequenced the genomes of 387 pairs of monozygotic twins and their parents, spouses, and children (Jonsson et al., 2021). They detected mutations that occurred during embryonic growth and found that the twins differed by an average of 5.2 mutations arising during early development. So, WGS can now reliably distinguish monozygotic twins, yet another case of technology development enhancing individual identification.

When it comes to technology, the criminal justice system is very conservative; novel DNA genotyping methods need to be validated in the lab and evaluated in admissibility hearings. In addition, the existing criminal databases all contain STR profiles, not SNPs or genome sequences, and the inexpensive standard STR genotyping is sufficient for most cases. For the next few years, the forensic implementation of NGS will probably be restricted to particularly difficult samples, like mixtures, and to complex and high profile cases.

Rapid DNA Technology

Unlike NGS, Rapid DNA instruments, capable of generating an STR profile within 90 minutes, do not offer a new strategy for genetic analysis with a dramatically increased Pd. Rapid DNA technology simply automates the standard STR genotyping by capillary electrophoresis. Nonetheless, the development and implementation of this technology promises to have a major impact on the role of DNA analysis in the criminal justice system. Several different companies have developed integrated microfluidic devices to automate the DNA extraction, PCR amplification of a panel of STR loci, capillary electrophoresis of the PCR products, and data analysis. The automation (basically, sample in, STR profile out) means that sophisticated forensic DNA analysis can now be achieved outside a laboratory, in places like police stations. The speed means that a profile can be generated and searched across a database while an arrestee is still at the booking station. The Rapid DNA Act of 2017 provided guidelines for the implementation of this new technology, and the FBI recommended analysis only of reference samples, typically cheek swabs and blood samples with lots of intact DNA. The performance of the instrument would, presumably, be less robust on the often degraded samples taken at the crime scene.

But the convenience and speed of a fully automated system the size of a microwave oven make using it on crime scene evidence and special cases of missing person identification pretty much irresistible, hence, inevitable. After the 2018 Camp Fire in Butte, California, many of the hundreds of victims could not be identified by traditional methods. The Rapid DNA company ANDE offered free DNA testing of the remains, and officials were able to bring closure to families of the missing much more quickly than they would otherwise. The use of Rapid DNA technology in the identification of missing persons does not raise many of the legal concerns that will attend its use in casework and its introduction in the courtroom. But the use of Rapid DNA instruments for casework has already started in some jurisdictions, and we can expect revised guidelines and requirements for additional validation studies to follow, um, rapidly.

Phenotypic and Biogeographic Ancestry Markers for Investigative Leads

Over the past few years, genetic markers of appearance, such as hair and eye color have been developed and applied to crime scene evidence. The association of genetic markers for these traits has been well established, with predictive values less than 100% but still potentially useful in investigations in the absence of eyewitness testimony. Genetic markers associated with other physical attributes like weight, height, and, recently, even age,[16] are being explored. Genetic markers indicating biogeographic ancestry can also be useful in guiding investigations. Although the vast majority of genetic variation is *within* populations rather than *between* populations, there are some

---

[16] Research by Steven Horvath has revealed that the patterns of chemical tags called methyl groups that bind to specific bases in genomic DNA in different tissues, like saliva or blood, can serve as a clock, indicating the age of the tissue.

genetic variants (alleles) whose frequency differs in different geographic groups, as we discuss in more detail in Chapter 13. These genetic variants, known as AIMs (ancestry informative markers) and used in the familiar commercial ancestry databases to provide estimates of ethnic ancestry, can also provide information about the ancestry of the individual who left the evidence at the crime scene.

These markers are all used to provide leads for the investigation rather than for calculating statistical metrics for individual identification, like the, by now familiar, RMP and LR estimates. But they, too, have their own probability estimates. In a suspect-less crime, a blood stain at the scene might, for example, be from a male individual with, say, a 75% chance of having blond hair, an 80% chance of blue eyes, a 90% chance of northern European ancestry, and a 75% chance of being over 50. These particular results might not be terribly helpful in Minnesota but they might be useful in a crime committed in the Four Corners area of the Southwest United States. The commercial laboratory Parabon has provided probability estimates of physical appearance to police investigators in several cases, including the Angie Dodge case, which we discuss in Chapter 6; and the forensic genetics company Verogen sells a PCR-based NGS kit with a panel of SNP markers for appearance and biogeographic ancestry.

The other recent significant development in forensic technology is not a new genetic marker or DNA technique but sophisticated and powerful software, known as "probabilistic genotyping," capable of analyzing crime scene evidence with multiple contributors and providing probability estimates that the suspect is included in the mixture. Forensic mixtures pose a unique set of problems, as we've seen in Chapter 4; probabilistic genotyping is discussed in more detail there.

The DNA technology and software available now for analysis of crime scene evidence is significantly more sophisticated and powerful than our initial tests in the late 1980s and early 1990s. With the exception of complex mixtures with trace DNA, current DNA technology is capable of delivering statistical metrics for individual identification that are truly astronomical and no longer as controversial as in the early contentious days of forensic DNA analysis.

6

Databases, Cold Hits, and Hot Button Issues

The idea of searching a DNA database for a "hit" has become so familiar that it's easy to forget how recent this practice really is. In October 1998, the FBI established the first national criminal DNA database, the National DNA Index System (NDIS). Shortly thereafter, in July 1999, the FBI reported its first "cold hit," linking six sexual assault cases in Washington, DC, with three sexual assault cases in Jacksonville, Florida, ultimately leading to the identification and conviction of Leon Dundas. As we all know, many criminal investigations have a victim and biological evidence from the likely perpetrator but, unfortunately, no suspect. In these cases, a genetic profile can be obtained from the forensic specimens at the crime scene and electronically compared to profiles listed in criminal DNA databases.[1] Such forensic data banks contain DNA extracts as well as computerized databases of coded DNA profiles of convicted offenders and, in some states, of arrestees. If the genetic profile of a forensic specimen turns out to match the profile of someone in the database, that individual may become the prime suspect in what was heretofore a suspect-less crime. Many of these crimes would never have been solved without use of such databases. In addition, in many forensic investigations, the *exclusion* of potential suspects whose DNA profiles are in an offender database but don't match the profile found at the crime scene can save valuable investigative time and resources as well as baseless suspicion visited on people in the database who are innocent of the crime. Data-mining of criminal databases can also provide crucial evidence in cases involving suspects who were wrongfully convicted. In many of these cases, although the genetic profiles of the evidence *did not* match the convicts in question, their exoneration took place *only* after a search of a criminal database identified someone whose profile did, as we saw in

---

[1] In principle, *any* DNA database could be searched for a match with the evidence or for a partial match, as is the case with "familial" searching.

Chapter 3. The crime scene evidence profile can also be uploaded into a forensic evidence database, so even in the absence of an identified suspect, matches between evidence DNA profiles from different crime scenes can indicate a serial offender. The power of searching convicted felon and, recently, other noncriminal databases for generating investigative leads is enormous, but the ethical issues and the social trade-offs raised by these searches are significant.[2]

Many states and many countries have compiled DNA databases consisting of genetic profiles obtained from blood samples or buccal swabs; initially these were from felons convicted of violent crimes, but the composition of these databases has since expanded in many jurisdictions. In the United States, CODIS (the Combined DNA Index System) is the FBI's program of support for criminal justice DNA databases as well as the software used to run these databases. NDIS refers to one part of CODIS, the collective national database containing the DNA profiles contributed by federal, state, and local participating forensic laboratories. In the 20 years between 1998, when both CODIS and NDIS were established, and August 2018, the NDIS database grew to contain 13,492,036 offender DNA profiles (~4% of the US population), 3,246,832 arrestee DNA profiles, and 879,945 DNA profiles of crime scene evidence.[3]

With over 431,850 "hits"—matches between forensic DNA evidence and particular DNA profiles in the database—reported by CODIS statistics during this 20-year period, searching such databases has clearly been very effective in identifying suspects and facilitating investigations, but arguments about the nature of the database (i.e., who should be included) and how the statistics of a cold hit match should be presented to the jury remain. Familial searching, looking for *partial* matches with the evidence profile, can point to relatives of the individual in the database and represents a recent, powerful, but still somewhat controversial use of existing criminal databases to identify potential suspects. This strategy raises many of the issues that characterize discussions of the criminal justice system in general, namely, trade-offs between civil liberties, privacy, equity, and effective law enforcement.

---

[2] The ethical issues raised by database searching are discussed in detail in Chapter 4 by Fredrick Bieber and Chapter 15 by Thomas White and Steven Lee in *Silent Witness* (Oxford University Press, 2021)

[3] https://www.fbi.gov/services/laboratory/biometric-analysis/codis/ndis-statistics.

What Is the Relevant Statistic for Matches
Resulting from Database Searches?

When a database search results in a "hit," a match between the evidence profile and someone in the database, what statistical metric should be presented to the jury? The answer could be the difference between an innocent or guilty verdict. Estimating the statistical significance of a match obtained with a database search should be straightforward but, for some reason, remains contentious. One view is that if the profile of a suspect identified by a database search matches the evidence, then the estimated population frequency of that particular genetic profile (equivalent to the random match probability, or RMP, in a nondatabase search case) is still the relevant statistic to be presented to the jury. The RMP estimates the probability that a randomly chosen individual in a given population would match the evidence profile. The RMP, as we've seen in Chapter 5, is estimated as the population frequency of the specific genetic profile, which is calculated by multiplying the probabilities of a match at each individual genetic marker (the "product rule"). An alternative view, often invoked by the defense, is that the size of the database matters and that number should be multiplied times the RMP. For example, if the RMP is 1/100 million and the database that was searched is 1 million, this perspective argues that the number 1/100 is the one that should be presented to the jury. This calculation, however represents the probability of getting a "hit" (match) with the database and *not* the probability of a coincidental match between the evidence and suspect (1/100 million), the more relevant metric for interpreting the probative significance of a DNA profile match. Although these arguments may seem arcane, the estimates that result from these different statistical metrics could be, as noted above, the difference between conviction and acquittal.

In "The Dark Side of DNA Databases," her October 2015 *Atlantic* article, New York University law professor Erin Murphy discusses the different metrics and the different numbers that could be presented to the jury. She cites a 2002 "cold hit" case in which a 70-year-old California man, John Plunkett, was accused of the sexual assault and fatal stabbing of a nurse in 1972, based on the results of searching a California state database. The prosecutor proposed presenting the RMP, which had been estimated as 1 in 1.1 million. The defense attorney argued that the "database match probability" of 1 in 3 (the database must have been around 350,000) should be the statistic presented to the jury. Professor Murphy suggests the use of yet another metric, known as the "n*p" statistic, namely, the estimation of how many of the men who lived in the area at the time of the killing and who were in the right age range would have also matched the crime scene evidence. This calculation involves multiplying the RMP (p) times the number of men in the right age range living in the area at that time (n, hence the "n*p" term). This number was two.[4] This line of reasoning would argue that the relevant metric to be presented to the jury is 1/2. However, the *probability* of a coincidental match is a more relevant metric than the *number* of men in a geographic area who might have the same profile, a calculation that is dependent on how "n" is defined. The judge in the Plunkett case ruled that the jury should hear only the RMP of 1 in 1.1 million. Plunkett was convicted of sexual assault and murder and is currently serving a sentence of life without parole.

My view, like that of the judge, is that the RMP is the most appropriate metric to represent the weight of the evidence. The other metrics measure different things. One approach that avoids passing judgment on the most relevant metric is to simply carry out additional DNA testing, whenever possible, when a database search results in a cold hit. For example, if a database search with 15 genetic markers generated a match, one could, in principle, then test an additional set of genetic markers to increase the already very high probability that the evidence was contributed by the individual identified by the initial search in the database . . . or exclude him. This approach was recommended by the National Research Council in 1992.[5] This strategy, however, fails to take full advantage of the set of 15 genetic markers used in the screening process and is unnecessarily conservative[6] and, arguably, simply unnecessary.

When testing with additional genetic markers is not possible (due to limited DNA evidence), as was the case with the semen sample in the Plunkett case, what statistic *should* be presented to the jury? Based on a recent California Supreme Court decision, the RMP or the population frequency estimate, calculated using the product rule, is the most relevant statistic. In *California vs. Nelson*, a case involving a cold hit match in 2002 between a profile from the 1976 murder scene of Ollie George, a 19-year-old college student in Sacramento, and the database profile of Dennis Nelson, an inmate

---

[4] The consideration of the "n*p" statistic is not unique to database searches and is equally applicable to the straightforward case of a match between a single source evidence sample and a suspect.

[5] National Research Council Report, *DNA Technology in Forensic Science*, 1992.

[6] "Conservative" in this forensic context means more favorable to the defendant, that is, a *higher* RMP.

at Folsom Prison,[7] the Supreme Court ruling cut through much of the confusion. This decision pointed out that these different statistics (e.g., RMP, database match probability), are all scientifically justified but that they address different questions. The defense in this case had argued that the RMP calculated by use of the product rule should be inadmissible if the suspect had been identified by a database search. The Court of Appeal opinion in this case, as well as an opinion from the highest court in the District of Columbia[8] that also addressed this precise issue, considered four different methods for calculating the statistical significance of a match.

The California Supreme Court ruled that the population frequency estimate (rarity), calculated using the product rule and equivalent to the RMP, is relevant even when the suspect is first located through a database search. The database match probability ascertains the probability of a match from searching a given database. As the Court noted, "*But the database is not on trial. Only the defendant is.*" Thus, the question of how probable it is that the *defendant*, not the database, is the source of the crime scene DNA remains relevant, and it is the rarity statistic that addresses this question. The 1996 recommendation of the National Research Council was interpreted by the FBI's Advisory Board to mean that that both the rarity statistic (i.e., the RMP) and the database match probability could and should be presented to the jury.[9]

Part of the confusion and the legal arguments surrounding the appropriate statistics for database "hits" stems from a 2001 study by state crime lab analyst Kathryn Troyer, in the DNA unit of Arizona's crime laboratory, cited by defense attorneys in the Plunkett case as well as in many others involving database searches. This study found that several samples from unrelated people in the database matched at many different genetic markers, an unexpected finding given the estimated rarity of these genetic profiles in the population. For some critics, this observation called into question the "cold hit" match statistics that were being presented by the prosecution. However, asking what is the probability that two samples within the database might match at

---

[7] Nelson was scheduled to be released from Folsom Prison one day before he was charged with the murder of Ollie George.

[8] See *U.S. v. Jenkins* (D.C. 2005) 887 A.2d 1013, 1019–1020 (*Jenkins*).

[9] The court's ruling also considered a statistic known as the Balding-Donnelly approach, which notes that, in obtaining a match in a database search, the other known profiles in the database have been eliminated so the probability that the person identified *is* the source of the evidence DNA profile is actually slightly *greater* than that suggested by the RMP estimate. While correct, this model, as noted by the Court, can be confusing to the jury (and possibly to the reader) and so it is the RMP that is typically presented to the jury.

multiple genetic markers is distinct from the question of the probability of a coincidental match for a particular genetic profile, which is addressed in the RMP estimation.[10] The unexpectedly high probability of any two samples in a database having the same profile reflects the very high number of pairwise comparisons (the profile of each sample is compared to the profile of every other sample) in this calculation, similar to what is known as the "birthday paradox." This problem can be expressed as how many people are required in a pool to have a more than 50% chance that any two of them would have the same birthday. This number turns out to be only 23. While this number *seems* unexpectedly low and this observation counterintuitive, considering the frequency with which multiple samples in a database might match seems, from my perspective, to obscure the issues of interpreting "cold hits," because it addresses the wrong question. Why does this misconception persist in discussions of forensic statistics, and why does this birthday paradox (or "parable" as Professor Murphy calls it) and the issue of multiple matches within a database continue to be invoked by defense attorneys? Many years ago, Upton Sinclair suggested a possible explanation: "It is difficult to get a man to understand something when his salary depends upon his not understanding it."

A more charitable interpretation is that some critics may have conflated the well-known statistical problem of multiple testing with the issue of database searching. In testing the hypothesis that a particular genetic variant is associated with a particular disease, we compare the frequency of that variant in a collection of patients (cases) to the frequency in a collection of healthy control individuals and calculate a metric known as the p value, based on the observed frequency differences between the two groups and the sample size. In these "association" studies and in statistical testing generally, the threshold for statistical significance has been set at a p value of <0.05, meaning that there is less than a 1/20 probability that the frequency differences we observed in the study are just due to chance and random "noise" and that the "null" hypothesis, no effect of the genetic variant, is, in fact, the correct one. But suppose we test 20 different genetic variants. Then, the probability that an apparent frequency difference for any one genetic variant might be observed simply by chance is very high and the threshold

---

[10] In a database search to generate an investigative lead, the profile of a crime scene sample is compared to every profile in the database—about 65,000 samples in the Arizona database in 2001. Troyer, however, compared all 65,000 profiles to one other, resulting in around 2 billion comparisons.

for statistical significance must be adjusted accordingly.[11] One such adjustment is the Bonferroni Correction, which simply multiplies the p value by the number of different genetic variants tested, similar to what some critics have proposed for calculating the probability metric in a database search. This confusion may be understandable, but hypothesis testing is a distinct statistical calculation from database searching and estimating the RPM.

Database Composition

From my perspective, the *California vs. Nelson* decision clarified the issues of which match statistic is relevant and admissible for cold hits generated from database searches. However, the *composition* of the databases that are appropriate to use remains controversial. The first criminal databases contained only profiles of felons convicted of violent crimes. Some states and many countries have expanded these databases to include genetic profiles of arrestees, a practice raising a variety of civil liberty issues. Some critics have argued that, since the criteria for arrest are subjective (e.g., probable cause), such databases are subject to bias and lack the more objective foundations of a convicted felon database (i.e., based on conviction). Others have noted that many of those arrested are never charged, raising issues of privacy and inappropriate governmental intrusion. Currently, 31 states are taking DNA from felony arrestees in addition to those convicted, whereas others do not. Among those states that do, some remove the profiles and destroy the DNA of arrestees who were never charged or found innocent, while in other states, these profiles remain part of the database.

The question of whether collection of DNA samples upon arrest, taken without a warrant and prior to a conviction, is constitutional under the Fourth Amendment, which prohibits unlawful search and seizure, was addressed by the US Supreme Court in *Maryland vs. King* in 2017. In an unusual 5–4 split (with Stephen Breyer joining the majority, conservative wing and Antonin

---

[11] The statistical issue of multiple testing, combined with recent advances in SNP genotyping, have influenced the way disease associations studies are now performed. Currently, so-called genome wide association studies are carried out with >500,000 genetic variants (SNPs) so the only way to reach statistical significance, given the necessary adjustments for multiple testing, is to design the study with very large sample numbers (e.g., >100,000 cases and controls). And the only way to achieve these very large numbers is to establish international consortia. Individual labs can no longer compete. In this way, statistical considerations have significantly transformed the sociology of science.

Scalia the liberal one) the Court upheld the Maryland DNA Collection Act (MDCA) and found that the warrantless collection of a cheek swab from Alonzo King in 2009 during booking did not violate King's constitutional rights. The MDCA allows state and local law enforcement officers to collect DNA samples from individuals who have been arrested for a violent crime or burglary or for attempted crimes of violence or burglary. While under arrest for first- and second-degree assault charges, King's DNA was collected and genotyped and the STR (short tandem repeat) profile logged into Maryland's DNA database. King's STR profile turned out to match DNA evidence from an unsolved 2003 rape, leading to his conviction for first-degree rape and a life sentence. The STR profile was the only evidence linking King to the rape. The trial judge denied King's motion to suppress the STR-profile evidence, but the Court of Appeals of Maryland reversed this decision, ruling that the DNA sample taken under the MDCA was an unconstitutional violation of King's privacy rights.

In the US Supreme Court's 5–4 decision to uphold the MDCA, the majority argued that the cheek swab was less intrusive than a blood draw and distinguished between the noncoding STR profiles and other potentially medically relevant DNA sequence information. In particular, they argued that obtaining an STR profile at booking, like the taking of fingerprints, was a form of identification for the arrested individual. Although this argument prevailed, many legal scholars were skeptical. "King was undeniably swabbed because police and prosecutors thought it would hit a cold case, not because anyone doubted Alonzo King was who he said he was," commented Jacob Sherkow on a Stanford Law School blog. Four of the Justices were also not convinced by the identification argument. As Scalia noted in his dissent, "DNA testing does not even begin until after arraignment and bail decisions are already made. The samples sit in storage for months and take weeks to test. When they are tested, they are checked against the Unsolved Crimes Collection—rather than the Convict and Arrestee Collection, which could be used to identify them" (Scalia, J., dissenting, slip op. at 12).

As we've seen in Chapter 5, the recent development and implementation of Rapid DNA technology, which is capable of converting a buccal swab into a genetic STR profile within 90 minutes, will certainly facilitate the expansion of arrestee DNA databases. Unlike the standard lab-based STR genotyping technology, Rapid DNA technology is actually fast enough that it *could* reasonably be considered a form of identification. So, in a sense, the emergence and deployment of Rapid DNA instruments in police stations may have rendered a not very convincing legal argument based on identification just a bit more plausible.

Population Databases and Ancestry

We've already discussed the composition of the criminal databases (convicted felons, arrestees, etc.) searched in suspect-less crimes as well as the statistical metrics used to present evidence to the jury. These statistics, such as the RMP, are based on the estimated frequency of a particular genetic profile in a reference population of randomly selected individuals. The population databases used for these calculations are distinct from the criminal databases used in the cold hit searches. Since the frequency of genetic profiles may be different in different ethnic groups, a variety of ethnic population databases have been established and used to estimate the RMP or other metrics (e.g., likelihood ratio) for interpreting the significance of a match.

The US forensic population databases are defined very broadly and based more on the convenience of federal census categories than on genetics: Caucasian, Hispanic, African American, Asian, Native American, Pacific Islander. The Hispanic category is particularly problematic in terms of genetics, as this group includes individuals with European, Native American, and African ancestry. How then, should the significance of a match be presented in the courtroom, given potential genetic profile frequency differences in these different population groups? One strategy is to provide the jury with RMPs calculated for different ethnic databases; for example, the RMP might be 1/500 million in one population and 1/100 million in another. It is assumed in the US criminal justice system that the most "conservative" estimate, the one most favorable to the defendant (i.e., the highest RMP), is the one that should be presented to the jury. Another strategy is to focus on the population groups present in the geographic area where the crime was committed.

In general, the STR forensic genetic markers used for individual identification have been chosen to *minimize* the frequency differences between populations so that the RMPs calculated from different population databases are not very different.[12] (If the genetic marker frequencies are similar in

---

[12] The sole exception is the Native American population. This population has less genetic diversity than the other US census group populations, so that RMPs calculated for this group may be higher than RMPs for other groups.

different populations, then the RMP estimates for a match between the evidence and suspect for the different populations will, accordingly, be more similar.) Why even consider population databases based on these ill-defined and self-identified racial groups? In 1996, the US National Research Council recommended that the RMPs be calculated using the product rule on allele frequencies from DNA profile databases that correspond to the suspect's racial background, when such information is available, based on the assumption that the most conservative RMPs can be estimated by using the suspect's own racial reference database.

Unlike the STR markers used in forensic databases, a small set of genetic markers whose frequencies *are* very different in different populations can be used to infer the ancestry of the individual who left a particular forensic sample. Although STRs not SNPs are used in forensic databases, using SNPs to illustrate this point is simpler. For example, a bi-allelic A/T SNP marker (where the A and T nucleotides are the two alleles present in the population for this single nucleotide polymorphism) could have a population frequency distribution of around 50% A and 50% T in many different populations. This genetic marker would be useful in individual identification because the three possible genotypes would all be relatively frequent (A/A and T/T at 25% and A/T at 50%). If, on the other hand, it had a population distribution of 99% A and 1% T in population 1 and 1% A and 99% T in population 2, this marker could be used to infer the population source of the sample.[13] Identifying the probable ancestry of an unknown perpetrator can serve as a valuable investigative lead. These markers are known as AIMS (ancestry informative markers). Commercial companies, such as Ancestry.com and 23andMe, use them in their genetic analysis to provide estimates of ancestry based on analysis of cheek swabs or saliva samples and comparison to various ethnic reference population databases.

Familial Searching

Genetic profiles derived from evidence in a suspect-less crime have been compared by law enforcement to the profiles in criminal databases for many years, as we've just seen. If a match is found, an investigation can be initiated

---

[13] These markers would *not* be useful in individual identification because in population 1, most (98%) people would have an AA genotype and in population 2, 98% would have a TT genotype.

and the individual whose DNA matched the evidence sample can, if still alive and a plausible suspect, be prosecuted. Typically, a cold hit means that the profile of the evidence matched the individual in the database at *all* of the genetic markers (STRs) used to genotype the contributors to the database.

When the search of a criminal database fails to reveal any cold hits (i.e., no complete matches at all genetic markers), broadening the search criteria to include *partial* matches (shared alleles) expands the search to relatives of individuals in the database. This is the investigative strategy now known as "familial searching," which involves scanning the database with specialized software to identify partial matches. Typically, a matching threshold (specifying the minimum number of genetic markers with shared alleles) is established, and the search might identify multiple individuals in the database whose profiles give a partial match to the evidence. This group can be further winnowed by a follow-up scan comparing the Y-chromosome profile of the evidence to the profiles of the database individuals who were identified in the partial match search. A match would identify a male relative as a potential source of the evidence. In these cases, the initial DNA data is used as an *investigative* tool rather than as evidence of identity to be presented to the jury. DNA data would be presented to the jury only if there is a match between the evidence and DNA of the suspect, initially identified by the partial match with someone in the database. The familial search strategy provides multiple "persons of interest" for investigative follow-up and, if a plausible case can be made for a particular male relative, his genetic profile can be obtained and compared to the evidence. In principle (and in practice in some cases), female relatives could also be identified if a genetic marker other than the Y-chromosome is used to follow-up on the initial partial match result.

Familial searching of criminal databases was first used in the United Kingdom in 2002 and, in 2006, was formally proposed in the United States by three distinguished forensic scientists, Frederick Bieber, Charles Brenner, and David Lazer, in an article titled "Finding Criminals throughDNA of Their Relatives." The familial searching strategy has subsequently resulted in identification and conviction of a suspect in many high-profile cases, notably the serial killer Lonnie Franklin Jr., known as the "Grim Sleeper" in California. Between 1985 and 1988 in Los Angeles, seven women and a man had been murdered and another victim had been raped but survived. After a hiatus of 13 years, the killings resumed, with a murder in each of three years, 2002, 2003, and 2007. There were no clear leads, but a common DNA profile in the crime scene evidence linked 6 of these 12 cases. No matches were found in searches of national and state databases. Subsequently, in 2008, the California Department of Justice carried out the first familial search in California but got no promising partial matches.

However, the California database, like others, is continuously expanding and, in 2010, after several hundred thousand additional profiles had been added, the LAPD requested another familial search. This scan revealed a partial match of the evidence with one of the database profiles that had been added in 2009. A follow-up scan of genetic markers on the Y-chromosome of the evidence and the individual identified in the database showed a match, suggesting that the evidence may have come from a male relative of this individual. The pattern of the partial match, namely that at each of the 15 sites used in the genetic testing, one of the two genetic variants (alleles) in the evidence sample matched the database profile, as well as the offender's likely age, pointed to the identified individual's father as the source of the evidence. Subsequent investigations revealed that the father, Lonnie Franklin Jr., had, in fact, lived near several of the crime scenes in Los Angeles at the time of the murders. In an attempt to recover a DNA sample, undercover policemen followed Franklin to a pizza parlor and collected napkins and a partially eaten slice of pizza. The genetic profile derived from the DNA extracted from these items matched the evidence samples, and Franklin was arrested and charged. In 2016, Franklin was found guilty of killing 9 women and a teenage girl over the course of 22 years and sentenced to death. He died in prison in 2020.

While acknowledging the societal value of these dramatic identifications, civil liberties organizations, the defense bar, and some academics have expressed objections to and concerns about familial searching. One concern has been that familial searching, by expanding the net of persons of interest based on those already convicted, would exacerbate the already existing racial inequities in our criminal justice system.

In a recent familial discussion of familial searching with my son Justin, he pointed out the considerable racial disparities in the California and federal criminal databases. He was very familiar with these issues, having worked as a Special Assistant Attorney General in the California Attorney General's office under Kamala Harris, where he helped develop Open Justice, a website and transparency initiative for the California criminal justice system. His concern was that familial searching, by turning relatives of those in the criminal database into potential suspects, would increase the already well-documented racial disparities in the criminal justice system. His point was indisputable, but I felt that eliminating familial searching would allow some criminals to evade justice, a bad outcome for *all* ethnic groups. And simply outlawing familial searching, as some states have done, does nothing to reduce the effects of systemic racism in our legal system.

Other objections remain. In various publications and presentations, Professor Erin Murphy has criticized familial searches for turning the fathers, sons, and brothers of convicted offenders into suspects. Her point that other kinds of DNA databases, such as those for missing persons, are *not* scanned for partial matches with the profiles from criminal evidence samples was, at the time, correct and to some, quite persuasive. Since then, however, the by-now familiar searches of genealogy databases have rendered this particular critique considerably less compelling. In addition, her characterization of all male close relatives as "suspects" seems overstated and ignores the role of the evidence, both genetic and investigational, in identifying an actual rather than a *potential* suspect.

Most discussions of this strategy are from the perspective of advocates, defense lawyers, or civil rights proponents or from law enforcement officers and prosecutors, and they tend to talk past one other. Legitimate concerns about civil liberties, privacy, and equity are often invoked to block the use of DNA analysis, and the representatives of law enforcement are often reluctant to acknowledge these very real issues. Discussions of familial searching typically invoke the concept of "trade-offs" between civil liberties and law enforcement. In some jurisdictions, the "trade-off" has been between two different controversial criminal database programs. In Maryland, for example, an arrestee database (albeit one specifying arraignment) was allowed but familial searching was outlawed.

Discussions of familial searching require considering the composition of the database to be searched. Many who would be comfortable with searching a convicted felon database might not be if the database instead consisted of people searching for relatives or ethnic origins (e.g., Ancestry.com). This difference reflects, rightly or wrongly, the sense that convicted felons have relinquished some fundamental rights. One of the other justifications, assumptions, and rationales for searching convicted offender databases for cold hits is that people previously convicted of crimes are more likely to commit another crime than the general population. An extension of this rationale in familial searching is that *relatives* of convicted felons are also more likely to commit crimes and therefore more likely to match the crime scene evidence than the general population. A recent Department of Justice (DOJ) study, for example, cited by Bieber, Brenner, and Lazar, reported that "46%

of jail inmates indicated they had at least one close relative who had been incarcerated." These rationales for familial searching of criminal databases suggest that searching an arrestee database compared to a convicted felon database is more problematic and less "efficient," since many of these individuals are innocent. Obviously, some people in a convicted offender database may also be innocent but, presumably, the proportion in an arrestee database would be higher. From a statistical perspective, the *prior probability* for a "hit" to an individual in a database containing a significant proportion of innocent people would presumably be less than in a convicted offender database. So, significantly less "bang for your buck." Also, an arrestee database, while arguably potentially valuable in solving suspect-less crimes, seems vulnerable to abuse, given the well-documented racial biases in the criminal justice system. Where to draw the line and the justification for its placement is not at all clear, as is the case with many discussions of societal "trade-offs."

Is it possible to define a set of conditions/constraints under which familial searching seems like an appropriate and justifiable strategy? Is trying to establish and observe these guidelines a fool's errand when the social desire to solve crimes is so strong and the potential of DNA analysis so powerful? A few years ago, the State of California made a good faith attempt, as reported by Michael Chamberlain, deputy attorney general with the California DOJ in "Familial DNA Searching, A Proponent's Perspective."[14] The guidelines suggest that familial searching should be initiated only for a specified set of serious crimes (e.g., murder, rape) and authorized by the state on a case-by-case basis. The databases to be scanned should be restricted to those of convicted felons; arrestee databases should be excluded from familial searching. A predetermined threshold for partial matching in the criminal database search should be established and a complete match of the Y-chromosome profile between the evidence and the individual in the database should be required to follow up on a "cold hit."

At a conference of the American Academy of Forensic Sciences in 2017, I had the opportunity to hear a panel of lawyers, scholars, and law enforcement professionals discuss the issues associated with familial searching. Erin Murphy and Michael Chamberlain were both on the panel. Toward the end of a lively and contentious discussion, Murphy stated her view that familial searching should *not* be permitted but conceded that, if it *were* to be done, it should be done with the constraints and protections of the California system,

---

[14] Chamberlain (2012).

as described by Chamberlain. Then, in 2018, the Golden State Killer case transformed the argument.

Genetic Genealogy

In the last few years, as we've all learned from the media, familial searching has moved from criminal to genealogical databases, as dramatically applied in the Golden State Killer case in 2018 and in many subsequent cases. This transition has completely upended the trade-offs and guidelines Chamberlain proposed as well as many of the arguments surrounding familial searches. Many of the rationales justifying familial searching of criminal databases, such as the recidivism rate and the presumed relinquishing of certain rights, do not apply to genealogical databases. Also, the concerns about racial disparities in criminal databases don't apply to these noncriminal databases either. In general it's very hard to draw lines in the sand when the sands are shifting and the technology is evolving. And it is particularly difficult when dramatic successes in identifying the perpetrators of truly heinous unsolved crimes are lauded in the media, making celebrities of the forensic scientists who carried out the complex genealogical analyses that finally led to the arrest of the Golden State Killer in 2018 and, shortly thereafter, to many others.

Golden State Killer Case

The Golden State Killer had committed 12 murders and was suspected of more than 50 rapes and over 100 residential burglaries in various parts of California from 1976 to 1986 (McNamara, 2018). These horrific crimes had remained unsolved for decades when, in 1994, a cold-case investigator, Paul Holes, working in the Contra Costa County District Attorney's office, picked up an old case file labeled the East Bay Area Rapist and began what would be a long and tortuous search with few, if any, leads. I had the opportunity to hear Holes speak at a forensic genetics conference in the fall of 2019 (International Symposium on Human Identification). His knowledge about the victims and the details of the crime scenes was exhaustive and his pursuit of the killer and commitment to solving this case was the unwavering focus of his career, continuing into retirement. Finally, in 2018, Holes, working with the results of a genetic genealogy search, identified Joseph D'Angelo, 72, as the Golden State Killer, 24 years after Holes had picked up that dusty file in the DA's office.

In 1994, when Holes first looked at the file, the only available DNA data was from the PCR-based HLA-DQalpha and Polymarker tests my colleagues and I had developed several years earlier. The results of these tests in the California case indicated that the same assailant was responsible for a series of sexual assaults in East Bay. In 1997, Holes discovered that the DNA analysis of several crime-scene semen samples from Southern California also had the same HLA-DQalpha and Polymarker genotypes. As the newer STR genotyping tests were applied to the evidence, it became clear that 10 different murders throughout the state could be attributed to a single serial murderer, now dubbed the Golden State Killer. Based on some non-DNA evidence from these serial break-ins, rapes, and murders, Holes started building a case against someone he considered a "prime suspect" but was disappointed and frustrated when that suspect's DNA profile did not match that of the semen stain found at one of the rape/murders. In his talk, Holes noted that this outcome reminded him that investigations need to be "led by the evidence not the suspect."

In 2018, Holes, with the help of Barbara Rae-Venter, a retired patent attorney and a genetic genealogy expert, decided to upload the DNA profile of evidence collected at the crime scene in Ventura where Lyman and Charlene Smith had been murdered by the Golden State Killer to an open-source public genealogy database, GEDmatch. Working with a small team, they carried out SNP genotyping[15] on the semen-sample DNA and found a partial match with several people in the database, including what would turn out to be a third cousin of DeAngelo, who had uploaded his DNA profile to this public database in an attempt to find his relatives. The partial matches allowed Holes and Venter to construct family trees using information from non-DNA genealogical records and focus the investigation on persons of interest within the assailant's approximate age range who were living in

---

[15] Criminal databases use the STR method of genotyping while most genealogy and ancestry databases use single nucleotide polymorphism (SNP) genotyping. DNA from crime scene evidence can be genotyped using SNP analysis and the SNP genotype can be loaded into a genealogy database to assess the degree of DNA sharing with people in the database. It addition, it is now possible using sophisticated statistical methods to infer a likely SNP genotype from STR data and vice versa. Once the suspect has been identified, however, and additional samples collected, the standard STR profile can be generated and compared to STR profiles of the evidence samples and RMPs calculated based on STR population databases. This result is what is presented to the jury.

California at the time of the crimes. DeAngelo, a former police officer who had been fired for shoplifting (dog repellent and a hammer, presumably used in the break-ins) and violent outbursts, met these criteria and became a suspect; the California authorities were able to collect his DNA surreptitiously from a so-called abandoned sample. The DNA profile matched the profile of the crime scene evidence that had been stored for decades. Following an extensive investigation, DeAngelo was arrested in August 2018, and charged with 13 murders. In August 2020, he pled guilty in exchange for a sentence of life without parole.

This collaboration of law enforcement with genealogy experts raised all the well-known concerns about familial searching *per se* but, in addition, about the ethics and privacy issues of screening noncriminal databases of people simply interested in finding relatives or learning more about their ancestry. From a civil liberties perspective, individuals who contribute to a genealogy database, unlike convicted felons, have not relinquished any rights and, from a sociological perspective, their relatives are not more likely than the general population to commit a crime. Searching commercial databases like Ancestry.com or 23andMe requires a court order, but GEDmatch was a public database to which people voluntarily uploaded their profiles and had much looser customer service agreements. Nonetheless, many critics believed that allowing law enforcement to search such databases was a violation of users' trust. Some of these concerns were captured in the title of the December 27, 2021, *New York Times Magazine* article, "Your DNA Test Could Send Your Relatives to Jail."

These ethical concerns notwithstanding, the success of familial searching of a genealogy database in the Golden State Killer case led very quickly to the application of this novel strategy to resolution of other long-standing unsolved cases. Less than one month after the arrest of Joseph DeAngelo in California, Canadian investigators, working closely with Washington State authorities, identified and arrested the alleged killer in the 30-year-old murder of Jay Cook and Tanya Van Cuylenborg, Canadian citizens, whose bodies had been found in an abandoned vehicle in Washington State. This identification was made possible by the work of CeCe Moore, a now famous genetic genealogist at Parabon NanoLabs, a DNA technology lab in Virginia. Moore had been working for many years on a strategy to combine DNA analysis with traditional techniques for finding adoptees' birth parents. In 2013, she founded DNA Detectives based on this strategy. The murder of Cook and Van Cuylenborg was her first criminal case.

A partial match of the profile from evidence on a blanket found at the crime scene with the profiles in a genealogy database allowed Moore to construct a family tree going back several generations. Tracing the contemporary descendants, on the branches of the family tree, who were in the area when the crime was committed identified a set of potential suspects, leading to the arrest and conviction of William Earl Talbott II, a construction worker and truck driver, in June 2019. (This conviction, the first for a defendant identified by genetic genealogy, was vacated in December 2021, due to juror bias. He will, presumably, be retried.) Moore is now the star of ABC's documentary, *The Genetic Detective.*

Angie Dodge Case: A Wrongful Conviction Overturned by Genetic Genealogy

A case that illustrates both the pitfalls and successes of genealogical searching as well as the all-too-familiar issues of coerced confessions and wrongful convictions is the Angie Dodge case. It also illustrates, as does the Golden State Killer case, the role persistence plays in achieving justice. Eighteen-year-old Angie Dodge was raped and murdered in Idaho Falls in 1996. Investigators collected semen and hair samples at the crime scene, and DNA analysis indicated that the samples came from the same unknown suspect. The local police investigated 20-year-old Christopher Tapp, but his DNA profile did not match the crime scene evidence. Nevertheless, after extensive questioning, Tapp confessed to the rape and murder and was convicted in May 1998. (In order to account for the DNA exclusion, the prosecutor's hypothesis included the supposition of another assailant in league with Tapp.) Although Tapp later attempted to appeal the conviction, the Idaho Supreme Court affirmed his conviction in 2001.

Chris Tapp's situation came to the attention of Greg Hampikian, a DNA expert and director and founder of the Idaho Innocence Project, who took on Tapp's case. In order to establish a new trial, new evidence would have to be introduced. In the hope of uncovering new data that might warrant a new trial, Hampikian suggested a familial search of the well-known genealogy database Ancestry.com. The initial identification of Michael Usry, a New Orleans filmmaker, as a person of interest through a search of this genealogy database by Idaho Falls law enforcement is an oft-cited example of familial searching gone awry. The crime scene evidence gave a match with

35 of 36 SNP genetic markers in the Y-chromosome profile of Usry's father, who had provided a buccal swab to a Mormon genealogy database, later acquired by Ancestry.com. In addition, it turned out that Michael Usry, the son, had been in Idaho at the time of the murder. Although he was ultimately excluded by additional DNA testing, this search was, arguably, problematic on many counts.

Genealogy databases typically do not use the same set of genetic markers (SNPs vs. STRs) and they are not subject to the guidelines, standards, and quality assurance (QA) procedures that characterize databases in the criminal justice system. Most importantly, however, unlike the discrimination power achieved by a match between many different autosomal genetic markers, a Y-chromosome match is not very discriminating and, in this case, only 35 of 36 Y-chromosome sites matched. Many other individuals, other than male relatives of Usry's father, would also show this degree of matching.

The database search failed to find a match for the profile of the semen sample found at the crime scene and, thus, no new evidence warranting a new trial. Tapp's attorney, Public Defender John Thomas, next contacted Judges for Justice, a national nonprofit organization that investigates false confessions, who found in 2014 that Tapp's confession was demonstrably false and coerced by police, but the Bonneville County prosecutor, Daniel Clark, continued to defend the murder conviction. After 20 years in prison, Tapp was finally released in March 2017 after accepting a plea deal with time served that dropped the rape charge but *retained the murder conviction.*

During the 1998 trial, Angie's mother, Carol Dodge, had herself become convinced of Tapp's innocence and, for the following two decades, remained committed to finding the real killer. In 2018, she contacted CeCe Moore and Parabon NanoLabs. The Idaho Falls Police Department had previously worked in 2017 with Parabon to develop predictions of physical traits based on analysis of crime scene DNA. The analysis in the Dodge case predicted an individual of northern European ancestry with fair skin, brown eyes, and brown or blond hair—all traits that were not very discriminating, particularly in Idaho.

At the same 2019 conference where I heard Paul Holes discuss the Golden State Killer case, I was able to hear CeCe Moore recount her experience with the Angie Dodge case. Moore initially told Carol Dodge that the DNA from the semen sample was very degraded and that generating a workable SNP profile might be so difficult that a database search was unlikely to be fruitful.

Dodge persisted and managed to convince Moore to submit a DNA profile generated from the semen specimen to GEDmatch, the public genetic genealogy database, for a familial search. As it turned out, several individuals in the database were identified who shared DNA regions with the DNA profile of the semen sample suggesting they were possibly distant relatives (second or third cousins) of the person, presumably the rapist and killer, who had left the evidence specimen.

Genetic genealogy involves constructing a large family tree going back several generations based on the individuals identified in the database search and on genealogical records. In terms of this book's metaphor and organizing principle, "genetic reconstruction of the past," this strategy goes back much further in time than the crime scene. Identifying several different individuals whose profile shares a region of DNA with the evidence profile allows a family tree to be "reconstructed." The shorter the shared region between two individuals or between the evidence and someone in the database, the more distant the relationship. This is because genetic recombination, the shuffling of DNA regions that occurs in each generation, reduces the length of shared DNA segments over time. So, in the construction of a family tree, the length of the shared region indicates how far back in time you have to go to locate the common ancestor.

Tracing the descendants in this family tree who were in the area when the crime was committed identifies a set of potential suspects. From the tree Moore constructed for the Dodge case, she eventually narrowed the list to six leads. One person of interest living in Twin Falls, Idaho, was excluded when the DNA profile of an abandoned sample (chewing tobacco) failed to match the semen sample profile. DNA from the other five also did not match the evidence profile. An obituary, however, pointed to another member of the family tree who had been missed. Moore and her team at Parabon were then able to identify another family member, Brian Dripps Sr., whose mother, the subject of the obituary, had been divorced and remarried. Dripps Sr. had been raised under his stepfather's surname and, at the time of the murder, had been living in Idaho Falls across the street from Angie Dodge's apartment. Investigators were able to collect a discarded cigarette butt; the DNA profile from the abandoned butt matched the profile of the semen and hair samples found at the crime scene. Brian Dripps Sr. was arrested and charged with murder and rape on May 15, 2019. He confessed and, on July 11, 2019, more than 21 years after the conviction, Prosecuting Attorney Daniel Clark filed a motion to exonerate Christopher Tapp.

Recently, CeCe Moore and Parabon Labs used genetic ancestry testing as well as genealogical analysis to identify a suspect in a homicide cold case in Lancaster County, Pennsylvania. In 1975, 19-year-old Lindy Sue Biechler was stabbed to death in her home. Analysis of the DNA extracted from Biechler's clothing pointed to a perpetrator with ancestral ties to a small town in southern Italy. The investigation then focused on the "few individuals living in Lancaster at that time of the crime that were the right age, gender, and had the family tree consistent with these origins," Moore said during the July 2022 news conference on the DNA-based resolution of this decades-old case. Eventually, David Sinopi, 68, who had lived in the same apartment complex as the victim became the prime suspect and Lancaster County law enforcement was able to obtain a DNA sample from a discarded coffee cup. Sinopi's profile matched the profile from Biechler's clothing, and he was charged with criminal homicide. Sinopi is currently being held without bail and awaiting trial.

Clearly, the familial searching of genealogy databases has proven enormously powerful. If some critics were concerned about searching criminal databases, the ethical and privacy issues raised by searching the now widespread and expanding DNA ancestry databases bring these concerns to a new and more urgent level. Over 15 million people have contributed their DNA to commercial databases like 23andMe or Ancestry.com to explore their genetic heritage or to find present-day relatives. (In 2022, my wife and I also sent a swab to Ancestry.com, a Christmas gift from my son Justin.) A 2018 study published in *Science* by Yaniv Erlich (no relative, as far as I know . . . but I have not consulted a genetic genealogist), the chief scientific officer at MyHeritage.com, a genetic ancestry company, estimates that using genealogic techniques, over 60% of Americans of northern European descent[16] could be identified based on the profiles only in the GEDmatch and MyHeritage databases. With the increase in these genealogy databases, the vast majority of Americans will soon be identifiable through the profiles of distant relatives.

Clearly, the guidelines established by the state of California for familial searching of criminal databases need to be updated for searching public, genomic, nonforensic databases. The value of searching these databases for

---

[16] This is the primary ethnic group using these databases. The searching of *criminal* databases has been criticized as exacerbating the existing racial disparities in the criminal justice system. Searching genealogy databases would not be subject to this particular critique.

ensuring public safety and bringing violent offenders to justice is indisputable. Concerns about privacy, civil liberties, and potential governmental intrusion, however, are very real. Given the potential to disclose private genetic information, affirmative informed consent to allow searches by law enforcement and genetic genealogists should be required by individuals contributing profiles to any of these databases. An "opt-in" for allowing searches by law enforcement rather than an "opt-out" for these databases is the new and current preferred policy model. This change, however, has consequences. As CeCe Moore noted in her 2019 lecture, if the GEDmatch database contained only those individuals who had currently "opted-in" at the time of her talk, she would not have been able to find Angie Dodge's killer. With the familiar trade-offs between civil liberties, privacy, and effective law-enforcement as well as the very substantial resources required to carry out genealogy-based strategies, these searches should be restricted to unsolved violent crimes without major investigative leads.

In the meantime, guidelines have not yet been established and searches continue. GEDmatch was launched by Curtis Rogers and John Olsen in 2010 as an open-source database for individuals searching for family members: they could upload their profiles generated by private companies like Ancestry.com and look for relatives based on partial profile matches. The Golden State Killer case in 2018 transformed the nature of these searches. By now (2022), the number of profiles in GEDmatch of those who have opted-in and are available for searching by law enforcement is over 500,000. Several other databases, like Family Tree, are also available for searches.

On December 9, 2019, GEDmatch, the formerly open access database, was acquired by Verogen, a California DNA forensics company, which had been spun out of the preeminent next-generation sequencing company, Illumina. This acquisition generated considerable concern and anxiety in the genealogical community. Brett Williams, the CEO of Verogen, has stated he will maintain the same opt-in standard for law enforcement searches that GEDmatch adopted in June 2019. In January 2023, the Dutch genomics company Qiagen acquired Verogen. Concern about the potential privacy abuses and actual costs (GEDmatch and Family Tree charge $700 per uploaded profile) associated with commercial ownership of genealogy databases prompted CeCe Moore and Margaret Press to launch a new non-profit DNA database *specifically to aid law enforcement*, the DNA Justice foundation. CeCe Moore's work at Parabon with existing databases has resulted in over 265 identifications in criminal cases; Margaret Press is the cofounder of the nonprofit DNA Doe project, a California organization that uses genetic genealogy to help identify crime victims. But concerns about privacy remain, whether the databases searched by law enforcement are commercial or nonprofit.

Some problematic searches go way beyond any potential violation of privacy. One of the most egregious abuses, reported recently (March 2022) by the San Francisco district attorney's office was a database search of genetic profiles obtained from rape kits. A woman was charged with an unrelated felony property crime based on analysis of DNA samples she had provided during a sexual assault investigation. The San Francisco Police Department had been using DNA from rape kits *intended to identify the rapist* for searches to identify suspects in other crimes, transforming victims into potential perpetrators. This practice creates obvious disincentives for rape victims to come forward, already an emotionally difficult undertaking, and ultimately damages the ability of law enforcement to solve crimes. According to the Justice Department's 2020 Criminal Victimization report, less than 23% of all rapes are reported to the police. In this case, the then San Francisco district attorney, Chesa Boudin, dropped all charges and proposed legislation to ban this practice. Police Chief Scott agreed, saying "We must never create disincentives for crime victims to cooperate with police" and expressed his willingness to work with the FBI in reviewing the department's practices as well as with federal officials, like Representative Adam Schiff, California, on legislation to protect DNA samples collected to solve crimes from inappropriate searches. Federal law already prohibits the inclusion of victim DNA in the national CODIS database. In response to the revelations of these searches by the San Francisco Police Department, State Senator Scott Weiner has sponsored Senate Bill 1228, Genetic Privacy for Sexual Assault Victims, that would prohibit including genetic profiles of victims into a database that could be searched for reasons unrelated to sexual assault. On April 26, 2022, Senate Bill 1228 was passed unanimously by the California State Senate Public Safety Committee.

The power of searching databases of all kinds combined with the inevitable and understandable pressure from law enforcement requires that clear and effective guidelines be established as soon as possible. Legislation at the federal and state levels will take longer but is no less necessary.

7

DNA, Doggerel, and Race Cards

The OJ Simpson Trial

The OJ Simpson trial, given its cultural significance on many different levels, warrants a chapter of its own. The TV reality-show-like courtroom proceeding gave rise around 20 years later to an actual 2016 TV drama, *American Crime Story: The People vs. O.J. Simpson*. For readers under 25 or from another planet, in 1994, the former football star, OJ Simpson was accused of the brutal stabbing of his ex-wife Nicole Brown Simpson and her friend Ron Goldman. The 1995 trial was televised and quickly became a major cultural event and social touchstone: "Did the Juice do it? "What would the jury decide?" were questions on everyone's mind. Of the many sobering lessons that can be learned from this seemingly endless trial (as noted by several wags, The Trial of the Century seemed to last for one), three stand out.

First, the American obsession with celebrity and the deep passions associated with racial injustice combined to obscure some of the critical issues and evidence and to bring about an unlikely, unexpected, and disturbing verdict. The jury seemed to be looking for a reason to find an attractive and famous black man innocent of two violent murders, and the defense team, with a little help from Mark Fuhrman, an LAPD detective whom the defense suggested had planted evidence, gave them several. The racist history of the LAPD gave them a few more. *This one's for Rodney King*[1] *and for all the other brothers oppressed by the system.*

Second, the other sobering, if not disturbing, aspect of the verdict is that it is entirely possible for a US jury to reject overwhelmingly conclusive scientific data. Was the jury's innocent verdict the result of inadequately clear presentation of the data and insufficient response to the defense hypotheses

---

[1] Rodney King was a motorist whose brutal beating in 1991 by four LAPD officers was videotaped. The footage was viewed all over the world. The officers were tried on charges of excessive force; three were acquitted, with no verdict for the fourth. Following the acquittals, the 1992 Los Angeles riots broke out and lasted for 6 days with 63 people killed and 2,373 people injured. After the riot, King's poignant plea "Can't we all just get along?" captured the spirit of a weary and shaken city.

of planted evidence by the prosecution? Or was it the inability of the jurors to understand the statistical significance of the observed matches for those specimens that could not have been planted, like some of the spattered blood stains?

Third is the pervasive influence of money in the US justice system. Is there any doubt that had the defendant been unable to hire his "Dream Team" of celebrity lawyers, the verdict would have been different? (As a basketball fan, I was skeptical that any team without Michael Jordan, Magic Johnson, or Larry Bird really deserved to be called the Dream Team—but the local celebrity-obsessed media loved the term—*it's LA, Jake*—).

I was scheduled to testify in the admissibility hearing, so I had the opportunity to review much of the DNA results. The evidence was overwhelming— a "slam dunk"[2] seemed to be the consensus among the forensic community. There were over 100 different evidence samples; genetic profiles of any one of them would have been dispositive. There were many small blood droplets on the walkway and gate at the Bundy (Nicole Simpson's residence) crime scene near the two bodies; all of these matched Simpson. (Apparently, he sustained a cut to one of his fingers.) Blood on the socks found at the foot of the bed at Simpson's house and the famous bloody gloves was also analyzed. The genetic profiles generated from these evidence samples matched either Nicole Simpson or Ron Goldman or were a mixture of both. The samples were analyzed by three different labs: Cellmark, a commercial lab; the LAPD lab; and the majority of the samples by the California Department of Justice lab. In 1994, the now standard analysis of short tandem repeat (STR) loci by capillary electrophoresis had not been implemented, so the evidence samples were genotyped with the PCR-based HLA-DQalpha and Polymarker test, and the D1S80 length polymorphism assay we had launched a few years earlier. Unfortunately, many of the crime scene samples, like the blood droplets on the walkway and gate, were not stored properly and the DNA was degraded, so that only the PCR-based assays but not the RFLP analysis could be carried out. Some samples had DNA that was sufficiently intact so that they could also be genotyped using RFLP analysis. Whatever genotyping method was used, however, the overall pattern of results was clear. The victims' blood was on samples at the OJ Simpson residence, and OJ Simpson's blood was present at the crime scene.

---

[2] This familiar basketball metaphor is actually quite misleading. For most people, a slam dunk is very hard. I could never dunk, even in my prime. An "uncontested lay-up" would be more apt.

The very large number of different forensic specimens that matched either the victims or the suspect and the complex pattern of mixed samples with profiles from both victims made the relevance and plausibility of an allegedly planted blood sample suspect. From my perspective, the lab genotyping results seemed to point overwhelmingly toward Simpson's guilt. As a scientist, I was dismayed that solid scientific evidence could be ignored. As a citizen, I was disturbed by dramatically different responses to the verdict by the African American and white communities. My scientist friends who were black, were, like me, taken aback by the verdict, but I was surprised and puzzled to see a few of my black acquaintances rejoicing in the verdict. Did the jury and much of the black community really believe he was innocent? Did they care or did their response simply reflect the pervasive distrust of the police that permeated so many black communities? The actor Courtney B. Vance, who played the defense attorney Johnnie Cochran in the 2016 FX series *American Crime Story: The People vs O.J. Simpson*, said that he celebrated the verdict back in 1995. "I cheered for Emmitt Till. I cheered for all the strange fruit that had hung on the tree for three centuries." The actor Sterling K. Brown, who played Christopher Darden, one of the two prosecuting attorneys, in the same series, sympathized with Darden's dilemma in being a black man prosecuting a black icon who had emerged from poverty to universal recognition and respect. Brown remembered thinking, as a 19-year-old, "Why are you trying to bring this brother down? This is the Juice!" Now, over 25 years after the trial, after Mike Brown, Tamir Rice, Laquan McDonald, Eric Garner, and so many other black men have died at the hands of the police and since the emergence of the Black Lives Matter movement in response to these deaths, it is clear, in retrospect, that the verdict and the response should not have been as much of a surprise as it was on October 3, 1995.

Two of the defense lawyers on the Dream Team, Peter Neufeld and Barry Scheck, had done and continue to do invaluable work in their immensely important Innocence Project. Since they both had a very good understanding of DNA analysis, I could only assume that they knew full well the implication of the DNA evidence to be presented in the trial. I have often wondered how Peter and Barry felt about their involvement in this trial. Did they take the case, in part, to get funds that could be used for the financial support of their righteous Innocence Project, a program that has done so much to advance the cause of justice?

The justification for the defense attorney's role in the judicial system is typically presented as functioning to assure a fair trial for their client and making sure that the prosecution does its job appropriately—fair enough—but, in practice, their goal seems, above all, to get their clients off. That is, after all, *how their job is defined.* Defense attorneys have told me that we can't really know the truth in a criminal investigation and so we just need to play out our roles assigned by the system. Ed Blake, who was on the defense team, expressed the view to me recently that this trial, as with all trials, is not about what actually happened but is about whether the prosecution met their burden of proof.

From my perspective as a scientist and legal outsider, there seems to be a fundamental asymmetry in the roles the prosecution and defense play in the justice system. The role of the prosecution, ideally, is to represent the People, aspiring to Truth and Justice. Although many prosecutors see their job as obtaining a conviction—i.e., *Winning*—their *assigned* role in the system is to help reach a just verdict. The prosecution is legally obligated to turn over any evidence that is potentially exculpatory to the defense. Deviations from this lofty ideal are, of course, not uncommon but they are, or should be, seen as malpractice, misconduct, or at least unprofessional behavior. The job of the defense, however, is much simpler—it is to represent the interests of the client, period. Dubious arguments or questionable data offered by the prosecution violate their proper role while, presented by the defense, they are, arguably, just part of the job.

The role the famous blood-stained gloves played in this trial illustrates this point. One glove had been found at the crime scene, while the other glove had been found by Mark Fuhrman at OJ Simpson's house. Ed Blake told me that he noticed that the Isotoner leather gloves were so saturated with Nicole Simpson's dried blood that they had become extremely stiff. He mentioned this observation to Barry Scheck, who then managed to bait the prosecution into asking OJ to put on the blood-stiffened gloves. According to some accounts, Marcia Clark was reluctant to have OJ attempt to put on the gloves, but Christopher Darden wanted to make the point that this critical piece of evidence truly belonged to Simpson. The resulting scene, with OJ trying unsuccessfully to place the recalcitrant gloves on his hands, became a dramatic turning point in the trial, immortalized in verse. Simpson was wearing latex gloves when he attempted to place the leather gloves on his hands and, because he had stopped taking his antiarthritis medication, his hands may have been slightly swollen, but the gloves, which normally stretched to accommodate the hand, didn't fit because they had been stiffened by Nicole

Simpson's blood. Had the prosecution attempted this ploy to try to convince the jury that someone's glove was not really his, this strategy might be seen as underhanded. For the defense, it was just good lawyering. In an attempt to recover from this debacle, the prosecution then argued that a leather glove, exposed to this amount of liquid, would be *expected* to shrink. As it turned out, Ed owned the very same kind of leather glove and the defense, following an experimental soaking, was able to show that the glove did not, in fact, shrink. Defense 2, Prosecution 0.

Many years ago, over a dinner at a Georgetown restaurant overlooking the historic canal, Peter Neufeld expressed to me the view that that since the system was corrupt and unfair, defending clients whether they were guilty or not was a righteous act. I don't know if he would still hold that same view now. I would like to think that, thanks in large part to the work of the Innocence Project, the justice system *is* now more just and fair than it used to be. Also, in this new era created by the capacity of DNA analysis to provide definitive evidence, the "truth," as defined by identifying the source of a biological sample, seems more accessible and, consequently, judicial outcomes should be more reliable. I think it's not naive to hope and expect that the injustices (wrongful convictions, false confessions, unreliable eyewitness accounts) that still are, sadly, part of our judicial system could be minimized as more information becomes available. In principle, DNA evidence should help attain that goal.

As it turned out, there was no admissibility hearing prior to the trial itself. Peter and Barry decided not to challenge the reliability of DNA technology *per se*, since it was so crucial to their Innocence Project work, stipulating to its admissibility and obviating my need to testify. They chose, instead, to base their case on the alleged mishandling of the evidence by the LAPD. In addition to celebrity lawyers, the Dream Team had some celebrity witnesses on their side. Although he never testified, my former Cetus colleague, Kary Mullis, who conceived of the polymerase chain reaction (PCR) and received the 1991 Nobel Prize in Chemistry for his idea, was waiting in the wings. He wrote in his memoir, *Dancing Naked in the Mind Field* (2000), that the DNA evidence was "botched" and should have been "thrown out on first principles," without providing any basis for these statements. In this media-saturated trial, Kary's testimony as a colorful, "maverick" Nobel laureate was eagerly anticipated by the media, and his memoir speculates about what he *might* have said when examined by Rockne Harmon, the tough and knowledgeable DNA expert attorney representing the prosecution. The defense team, however, wisely chose not to put him on the stand.[3]

My friend and colleague, Ed Blake, less of a media celebrity but much more knowledgeable about forensic DNA analysis, was also on the defense team. A friend of Peter and Barry, Ed told me he thought the defense team had thoroughly outmaneuvered the prosecution, forcing them on the defensive on a number of issues, and was very critical of the prosecution's responses and overall strategy. Ed played an invaluable part in ensuring that the actual lab work done by Gary Sims of the California Department of Justice lab was reliable, but he was never called to the stand to testify. He was critical of the way some of the evidence samples had been collected and stored but felt that the PCR-based genotyping met his rigorous standards. I suspect he thought that OJ probably committed the murders but that the prosecution didn't meet their burden of proof and, thus, didn't really "deserve" to win.

Another celebrity, Henry Lee, a knowledgeable and engaging forensic scientist, was also on the defense team. A well-known expert witness and popular lecturer, now with his own TV show, *Trace Evidence: The Case Files of Dr. Henry Lee*, Lee reveled in lurid presentations full of grisly photos of crime scenes. His lectures were a bit much for me, but audiences loved him. He introduced one of the many red herrings that further confused the deliberations in this already complicated case, namely the idea that the detection of a chemical blood preservative, EDTA, in the crime scene blood sample would demonstrate that the blood had been planted. When blood is collected in a clinical setting or, as here, for a reference sample in the context of a forensic DNA analysis, the collection tube often contains the chemical EDTA to prevent coagulation and to serve as a preservative. So, it's logical to assume that the finding of EDTA in a crime scene blood sample would be consistent with the blood having come from a collection tube rather than from a human body. Thus, detecting EDTA could, *in principle*, support the defense claim that the blood evidence had been planted. The issue, however, is one of *concentration*. There are very low levels of background EDTA in the environment while the concentration of EDTA in a blood collection tube is very high. The levels detected were low and not really consistent with the

---

[3] To anyone who has read Mullis's memoir, his interview in *Esquire*, or his testimony about DNA evidence in other trials, the wisdom of this strategic decision would be apparent. His credibility would have inevitably been attacked in cross-examination based on some of his indefensible views, such as HIV is not the cause of AIDS. Sadly, this position, articulated by a Nobel laureate, informed the public health policy of South Africa, with disastrous consequences.

claim of planted evidence, but, nonetheless, this issue created a lot of court-room "smoke" and the jury may well have imagined a "fire."

Many of the episodes in the popular reality TV series that the trial ultimately became did, however, serve to educate the public on the role of DNA analysis in criminal investigation and trials. The general expectation that DNA evidence should be part of criminal investigations, whether real or fictional, that emerged from the Simpson trial is probably a positive development for the criminal justice system. Certainly, the screenwriters for the burgeoning CSI franchise owe a huge debt to this televised trial of the century. The other media consequences are also considerable; in the wake of the OJ trial, a cable network (Court TV) was launched as well as the careers of several now ubiquitous legal talking heads on cable news.

Finally, on a more positive note, the verdict reinforces the eternal power of poetry. Is there a more consequential couplet in the history of verse than "If it doesn't fit, you must acquit"? What if the prosecution doggerel team had responded with, "If the genotypes match, then you must dispatch"? We'll never know.

Looking back on this trial and its outcome, which was puzzling and deeply troubling at the time, from the perspective of 2022, following the murders of George Floyd, Brionna Taylor, and too many others, the jury's verdict just seems deeply sad. The willful ignorance of scientific evidence remains a feature of contemporary political discourse, and the systemic racism that helped shape the jury's response to the evidence, is, sadly still with us.

8

Closure and Justice

Identifying the Missing

On October 18, 1984, Laura Bradbury, a three-year-old girl, went missing from her family's campsite in Joshua Tree National Park. A massive search with 250 people and a bloodhound failed to find any evidence related to her disappearance. Three days later, the search was called off, but the Bradbury family mobilized a national effort, clinging to the hope that Laura had been kidnapped but was still alive. As a symbol of the national campaign to find missing children, Laura was one of the first children to have her image displayed on milk cartons.

In March 1986, a skull and other remains were discovered by hikers, only five miles from the family's campsite at the Indian Cove Campground. The skull bone fragments appeared to be those of a young child. The initial attempt at DNA genotyping by RFLP proved inconclusive but, in 1990, Mark Stoneking, a visiting scientist in my lab at Cetus, was able to get access to the DNA from the skullcap from the San Bernardino County Sheriff's office. Mark had just developed a PCR-based test for analyzing polymorphism in mitochondrial DNA, working with other members of the lab, notably Russ Higuchi, who had helped develop the HLA-DQalpha test (Stoneking et al., 1991). When Mark analyzed the sample using a probe-based technology similar to our HLA-DQalpha test, the mitochondrial genotype of the bone fragment turned out to match that of Patty Bradbury, Laura's mother, indicating with a very high probability (>99%) that these remains were from Laura Bradbury. The results of this first DNA identification of a missing person were, of course, devastating to the Bradburys, who had hoped Laura might be found alive, but were, with time, reluctantly accepted, and ultimately provided some degree of closure to the grieving family. The perpetrator was never found.

Every missing person case is a unique and heartbreaking story, whether it's an individual skeleton, the mass graves of the Balkan wars or the Rwandan genocide, or the remains of the victims of the 9/11 World

Trade Center terrorist attack. The forensic strategies for exhuming and identifying these human remains, however, are essentially the same. DNA is extracted from bones or teeth and a genetic profile is generated; this profile is then compared to the profiles of a relative or of samples (e.g., hair from a hairbrush) from the missing person, if available. Identification can be based on a *complete* match between the mitochondrial DNA of the remains and a maternal relative. Alternatively, it can be based on a *partial* match of the short tandem repeat (STR) profiles between the remains and a relative. The profile of the remains can also be used to search a missing persons database. While the DNA technology has evolved since this first identification, transitioning from probe-based mitochondrial DNA genotyping to STR profiling to next generation sequencing (NGS), the overall strategy for identification remains the same. Some of the improvements in the technology used to identify human remains owe a lot to recent developments in ancient DNA analysis, in particular, the improved methods of DNA extraction and NGS library preparation for samples with highly degraded DNA.

It is not just the methods that have moved from evolutionary DNA studies to missing person cases but, in many cases, the scientists themselves. Mark Stoneking, whose work is discussed in several chapters of Part II ("Reconstructing the Ancient Past"), was one of the scientists in Allan Wilson's lab, along with his wife, Linda Vigilant, who contributed to the Out of Africa hypothesis, based on analysis of mitochondrial DNA. He worked on the Bradbury case as well as several criminal cases after he went to Pennsylvania State University, eventually moving from forensics back to evolutionary studies at the Max Planck Institute for Evolutionary Anthropology at Leipzig, the Mecca of ancient DNA studies. Linda was the main reason Mark came to our lab. Having finished his postdoctoral fellowship in the Wilson lab, Mark decided he needed to find a spot in a SF Bay area lab while he waited for Linda to complete her PhD at UC Berkeley. So our lab, and the field of forensic genetics, were the grateful beneficiaries of this decision to support his wife's career.

When reference samples are unavailable, the strategy of genetic genealogy, which proved so valuable in the search for the Golden State Killer and other murderers, can also be applied to identification of the missing. In December 1988, the body of a young woman who had been strangled was found on Interstate 59 in north Georgia. In March 2022, this strategy of searching genealogy databases for a partial match with the profile of the unidentified remains finally identified the murder victim as Stacey Lyn Chaborski, a 19-year-old who had gone missing during a hitch-hiking trip across the United States. A few months later, the murderer in this case was also revealed by genetic genealogy to be a truck driver who regularly drove the Interstate 59 corridor. Bodily fluid found close to Ms. Chahorski's body was eventually traced to Henry Frederick Wise, who had died in a stunt car accident in 1999. Mr. Wise had a criminal history that included theft and assault. This case was the first time *both* the victim and her murderer were identified by genetic genealogy, which was carried out by a private Texas-based DNA company, Othram, working closely with the Georgia Bureau of Investigation and the FBI.

The Celebrity Missing: Identifying the Remains of Historical Figures

DNA testing has also been applied to the identification of famous historical figures, the "celebrity missing." Unlike the massive efforts to match profiles from mass graves with a huge database, these investigations start with a limited, binary question. . . do the bone fragments found under a parking lot in Leicester, England, come from King Richard III . . . or not? Richard III died of wounds incurred at the Battle of Bosworth on August 22, 1485. Historical records indicate that his remains were buried in the Grey Friars church in Leicester. In 2012, a skeleton was excavated underneath a parking lot thought to lie above the presumed site of the Grey Friars friary. Radiocarbon dating data were consistent with the bones being his remains. Since Richard III had no children, the only potential reference samples were matrilineal and patrilineal relatives. Using NGS, the mitochondrial genome sequences derived from these remains were compared to those of two different living matrilineal relatives (King et al., 2014). No differences were found with one relative and only one single base substitution with the other relative, indicating with a very high degree of probability that the remains were from Richard III. It was not unexpected that a single mutation in the mitochondrial DNA sequence would have occurred in the many generations since 1485. What was unexpected were the results of comparing Y-chromosome profiles from the bone fragments with five different patrilineal relatives, descended from the 5th Duke of Beaufort (1744–1803). The Y-chromosome profiles from the remains did *not* match those of these distant male-line relatives, due, presumably to a false paternity issue sometime in the 24–26 generations since 1485. Since other false paternity events can be detected in the genetic and genealogical records, this observation, while "unexpected" given the mitochondrial DNA match, is not so surprising.

The first "celebrity relic" identification and an important milestone in forensic DNA testing was the analysis of mitochondrial DNA generated from bone fragments thought to be from the last imperial family of Russia, the Romanovs, executed on July 16, 1918, by Bolshevik troops. In 1979, the mass grave of Tsar Nicolas II, his wife Tsarina Alexandra, and three of their daughters, along with some of their servants, was discovered near Yekaterinburg,[1] east of the Ural mountains. Following the fall of the Soviet Union, their remains were exhumed for DNA testing. In 1994, an international team led by Peter Gill of the UK Forensic Science Service and Pavel Ivanov, a Russian geneticist from the Engelhardt Institute in Moscow, compared the mitochondrial profiles generated from these remains to the profiles of maternal relatives of the Romanovs (Gill et al., 1994). They first determined the presence of five females and four males in the grave, based on DNA gender testing, and, using the newly developed STR genotyping markers, identified a father and mother, along with three daughters. The son Alexei and one daughter were missing and the other four remains were assumed to be the servants. The mitochondrial DNA sequences derived from the presumptive skeletal remains of Tsarina Alexandra and her daughters matched those generated from blood samples of Britain's Prince Philip (Tsarina Alexandra's grandnephew). Tsar Nicolas II did not share this maternal lineage, so the remains of his brother, the Grand Duke George, were exhumed. The mitochondrial DNA profiles derived from the Grand Duke George's remains matched those of the presumptive remains of the tsar. A rare heteroplasmic site, C/T at position 16,169, was found in both sequences. At the time, heteroplasmy, the presence of two mitochondrial DNA sequences that differ at a single base within an individual, was not well understood. This feature of mitochondrial DNA sequences is discussed in more detail in the discussion of mixtures in Chapter 4.

---

[1] Yekaterinburg, the fourth-largest city in Russia, is the home of the golden-dome Church of the Blood built on the site of the Romanov executions. It is also the home of the professional basketball team that Brittney Griner, the WBNA basketball star, played for before she was convicted in a Moscow court of cannabis possession in February 2022 and sentenced to nine years. She languished in a Russian penal colony until she was exchanged for the Russian arms dealer Viktor Bout in December 2022.

Despite the convincing evidence establishing the identify of these remains, questions and doubts were raised by some scientists,[2] leading to the Russian Orthodox Church's refusal to accept the DNA results and reluctance to give the remains, interred in Saint Petersburg, full burial rites. In 2000, the tsar and his family were canonized and the church wanted additional assurances that the identification of these bones, now holy relics, was firmly established. In 2007, a grave containing the two missing Romanov children was discovered around 70 meters from the larger burial site. Testing these remains, and retesting the original material, with the improved STR genotyping methods as well as mitochondrial DNA sequence analysis, provided convincing evidence that the remains are of the two missing Romanov children, Alexei and one of his sisters, in support of the milestone identification of the Romanovs by Gill and colleagues (Coble et al., 2009). The final resting place of the Romanov remains is now up to the Russian Orthodox church.

Mitochondrial DNA testing was also used in the search for the bones of Sister Marija Krucafiksa Kozulic, a nun from the port town of Rijeka in northern Croatia, who died in 1922. She was buried, along with her biological sister Tereza and the comingled remains of other nuns from the Society of Sisters of the Sacred Heart of Jesus, in a tomb in Rijeka. In recognition of a virtuous life committed to helping the poor and less fortunate, Sister Marija is currently being considered for sainthood by the Vatican. Since canonization involves the reburial of the remains in a sacred site so that they can be properly venerated, this process requires the identification and transfer of her remains. Soon after the exhumation of skeletal remains in December 2011, Dragan Primorac, the charismatic scientist/physician and former minister of science, education, and sports of the Republic of Croatia, began to organize an international effort to try to identify Sister Marija's remains. The long bones were sent from the Rijeka University to the Clinical Hospital Center at Split, where they were measured and prepared for DNA extraction. I became friends with Dragan when he spent some time in the early 2000s working with my colleague Sandy Calloway and me on our recently developed probe-based genotyping technology for the hypervariable region of mitochondrial DNA. During that time, Dragan was pioneering the DNA identification of skeletal remains in Croatian and Bosnian mass graves.

---

[2] A publication by Alec Knight (2004), a Stanford geneticist, challenged the findings of the Gill et al. paper, arguing that the results were likely due to contamination. These concerns were rebutted by two articles in *Science* (Hofreiter et al., 2004, and Gill and Hagelberg, 2004).

Confronted with the highly degraded genomic DNA extracted from these samples, Dragan established a collaboration with Sandy to see if the probe-capture/NGS-based technology we had just developed for full mitochondrial genome sequencing, described in Chapters 4 and 5, might prove valuable for this effort.

Sandy, seven months pregnant, traveled to Croatia in December 2013; her graduate students, Sarah Copeland and Cassandra Taylor, were eventually able to generate full mitochondrial genome sequences from some of these highly degraded DNA samples, an impressive technical feat. However, at the time, there was no reference sample available for comparison. In November 2018, a team of forensic geneticists, well versed in the complexities of mitochondrial DNA analysis and led by Mitch Holland (Pennsylvania State University) and Charla Marshall (Armed Forces Medical Examiner System) used similar NGS technology to obtain mitochondrial genome sequences from a variety of femoral samples from the Rijeka tomb. The results revealed two right and two left femurs that had the same mitochondrial sequence and proved to be the only two maternal relatives found in the burial site. The sequences shared a rare heteroplasmic site; due to the rarity of the mitochondrial sequence, it was assumed that these might be the remains of the Kozulic sisters, Marija and Tereza. A buccal swab from a known paternal niece, now deceased, provided a reference sample for comparing autosomal STR and SNP profiles. Although only partial SNP and STR profiles could be generated from the remains due to extensive DNA degradation, the comparison with this paternal relative provided strong statistical support that these remains were indeed those of the Kozulic sisters (Marshall et al., 2020). It is not possible, however, in the absence of additional reference samples, to distinguish which remains belong to which sister. If the Vatican beatification process results in sainthood for Sister Marija, her sister Tereza may find a sacred resting place as well.

Armed Conflicts and the Missing

Unlike the identification of the remains of individual missing persons, like Laura Bradbury or of historical figures, i.e., projects with a specific hypothesis, the identification of the missing in violent conflicts and mass graves is much more difficult. The same genetic techniques are used, but, due to their massive scale, these projects require a very high level of organizational and administrative structure. The comingling of remains creates additional difficulties. The introduction of DNA technology was critical in this effort, given the difficulty of identifying remains on this scale based on traditional methods. The goals of forensic scientists are to (1) identify the remains of the missing for return to families for proper burial, (2) gather physical evidence for prosecutions, and (3) establish the historical record of what happened. Unlike DNA analysis, traditional forensic science also seeks to determine the manner and cause of death, a necessary element in some prosecutions.

The leading organization in this field is the International Commission on Missing Persons (ICMP), created in 1996 at the G7 summit conference in Lyon, France, at the instigation of US President Bill Clinton, to help account for the many thousands of people who had been missing as a consequence of the wars in the former Yugoslavia.[3] In 2003, the ICMP's mandate was extended to address the problem of missing persons globally, including after natural disasters, like the Indian Ocean tsunami that devastated southeast Asia just a year later. Initially, the ICMP was based in Sarajevo, befitting its focus on the former Yugoslavia. In 2008 my colleague Sandy Calloway traveled to Sarajevo to help train the DNA lab on the PCR-based mitochondrial test we had developed, based on Mark Stoneking's original work on the Bradbury case many years earlier. DNA testing on degraded skeletal remains had, for many years, relied on the analysis of mitochondrial DNA polymorphism because, as we've seen in previous chapters, there are many more copies per cell (around 1,000) than nuclear gene copies (2), increasing the probability of a successful PCR amplification from the specimen. Although the discrimination power of mitochondrial DNA is less than that of a panel of nuclear genetic markers, mitochondrial DNA genotyping could still be very powerful for identifying missing persons when combined with additional non-DNA information (medical and dental records, fingerprints, clothing, etc.),[4] particularly with a limited number of possibilities.

This strategy proved to be very effective for testing presumptive hypotheses of identity for confirmation. The massive scale of the death toll in the Balkan wars (an estimated 40,000 missing), however, precluded the effective use of mitochondrial DNA typing for identifying the remains, many of which were buried in mass graves with extensive comingling of skeletal bone fragments.

[3] The ICMP is funded by financial contributions from governments, foundations, corporations, and individuals.
[4] Fingerprints are not, of course that helpful for identifying skeletal remains.

The victims of the Srebrenica massacre, over 8,000 Bosniak Muslim men and boys killed by Bosnian Serb forces, were initially buried in mass graves but then, in an attempt to avoid accountability, exhumed by the Serbian perpetrators and redistributed to 90 secondary sites, further fragmenting the already degraded skeletal remains. For a project of this magnitude and complexity, the resolving power of mitochondrial DNA might not be sufficient, especially in the absence of medical or dental records. Over time, the ICMP developed an STR genotyping system with a high throughput laboratory workflow, a centralized DNA database, and a kinship software for database searching and carried out a massive effort to collect reference samples from relatives for each missing person for the DNA database. The ICMP also had to build relationships and establish trust with the families, many of whom were skeptical of international organizations after the failure of UN peacekeepers to prevent the violence and killings that took place in Srebrenica.[5]

Tom Parsons, who became the director of science and technology for the ICMP in 2006, oversaw the implementation and expansion of this highly sophisticated system of STR profiling and database searching to identify remains. A critical development was the use of new and more sensitive methods for extracting DNA. Nonetheless, some samples did not yield useful STR profiles due to extensive DNA degradation. As we've seen in Chapter 5, determining the number of tandem repeats, which is the basis of STR genotyping, requires the PCR amplification of DNA fragments that are long enough to span the entire tandem repeat region. Recently, the ICMP initiated an NGS system capable of generating a huge amount of data for SNP polymorphisms, a strategy ideally suited for the analysis of the degraded DNA samples that failed STR profiling. The massively parallel property of NGS (also known in forensic circles as massively parallel sequencing, or MPS), provides so much genetic information that kinship relationships can be established even with distantly related relatives. Over the course of many international conferences, Tom and I had many lively discussions about the relative merits of different genetic technologies and their application to the challenges that the ICMP faced. My view was that sequencing the entire mitochondrial DNA genome using NGS technology, not just the most polymorphic segment, the so-called hypervariable region, had the potential to provide valuable data for the identification process, particularly, for degraded samples, provided

---

[5] The UN peacekeepers were Dutch and the International Criminal Court is located in The Hague, where the ICMP later moved, as discussed below.

that the population database of mitochondrial genomes would be expanded. Since estimating the random match probability (RMP) for mitochondrial DNA sequences cannot use the product rule, as we've seen in Chapter 5, the discrimination power is determined by the size and composition of the database.

In 2016, the ICMP relocated to The Hague, as part of its transition to the status of a truly international organization, where it established new, state-of-the-art laboratories, complementing the work of the laboratories that remained active in Bosnia. Since its beginnings, the work of the ICMP has resulted in the identification of 27,000 of the approximately 40,000 persons reported missing as a consequence of the armed conflict in the former Yugoslavia. This number includes the 8,500 missing persons identified by traditional methods prior to the widespread implementation of their DNA testing program. There are still thousands who remain unidentified, so the work continues. In a world with devastating armed conflicts, violent repression, refugee and migration crises, and natural disasters, the mission of the ICMP has become truly global and more important than ever. It has also taken on some of the international legal and human rights issues raised by these conflicts.

In addition to its humanitarian mission of returning the remains of the victims of massacres and armed conflicts to their families, the ICMP has been instrumental in providing evidence and building cases in courts and tribunals against those responsible for these war crimes. The testimony of ICMP experts at the International Criminal Tribunal for the former Yugoslavia as well at the War Crimes Chamber of the Court of Bosnia and Herzegovina in Sarajevo was critical in a number of cases, including that of the first president of Republika Srpska, Radovan Karadzic, who was convicted in 2016 and sentenced to life imprisonment for genocide and crimes against humanity, including the Srebrenica massacre. The notorious Bosnian Serb general Ratko Mladic was convicted of these same crimes the following year.

The legal requirements to generate the genetic results appropriate for courtroom evidence are sometimes in conflict with the commitment to work closely with the families to ensure that as many of the remains are identified and returned as possible. In 2011, the ICMP commissioned a countrywide opinion survey asking whether the respondents would be comfortable supplying their genetic and personal data to the courts as evidence in war crimes, crimes against humanity, and genocide. Only around 58% said yes. A few years later, the ICMP asked the consent of over 1,200 relatives of those missing in the Srebrenica massacre to submit their personal genetic data to the trial chamber in the case of Radovan Karadzic. Unlike the respondents to the countrywide survey which addressed *all* relatives of the missing, these families had all been informed by domestic courts that their missing relatives had been found and identified by DNA matching. Fewer than 1% of these relatives declined to provide written consent. This testifies that the families of the missing seek "not only closure but also justice" (Andreas Kleiser and Tom Parsons, in Erlich et al., 2020). The experience and expertise of the ICMP in balancing these sometimes competing claims can be valuable as other smaller organizations address this problem.

Although the ICMP is by far the largest organization using genetic information to address the issue of missing persons, they were not the first. In Argentina, the pioneering work of a large group of forensic scientists, forensic anthropologists, activists, and family members established the guidelines for all subsequent efforts to identify the missing. During the military dictatorship in Argentina that began with the overthrow of President Isabel Peron in 1976, thousands of political activists and opponents of the newly installed military junta, headed by General Jorge Rafael Videla, were brutally murdered. Death squad members, known as *grupos de tareas*, as well as regular military units detained student activists, trade unionists, and other "left wing subversives," torturing and killing many of their victims and burying them in unmarked graves, or dropping them from planes into the ocean (National Commission on the Disappeared, 1987).

The children of the "disappeared," and children of women who gave birth in detention, were systematically taken away and put up for adoption by the regime, using falsified papers. These orphaned children were given to childless families in the military and police as well as to friends of the regime. In April 1977, a small group of middle-aged and elderly women gathered in the Plaza de Mayo across from the presidential palace in Buenos Aires to demonstrate their opposition to the mass killings carried out by the junta. The women, who became known as the *Madres de Plaza de Mayo*, wore white kerchiefs embroidered with the names of their disappeared children and marched silently in a circle, in defiance of the junta's ruling against public gatherings and demonstrations. In October 1977, they were joined in their weekly vigil by another group, the *Abuelas de Plaza de Mayo*, led by an art history teacher, Maria Isabel Chorobik de Mariani, and focused on finding their kidnapped grandchildren and ultimately restoring them to their biological families.

In 1982, the junta, weakened by their loss in the war with the United Kingdom over the Falkland Islands (or *Islas Malvinas*) and facing mounting opposition, lifted the ban on political activity and promised elections in October 1983. After six years of military rule, Raul Alfonsin, a civil engineer and leader of the Radical Party, was elected president. He ordered the prosecution of members of the military junta and established the National Commission on the Disappearance of Persons, chaired by the novelist Ernesto Sabato, to investigate what had happened to the disappeared and to establish accountability for these crimes.

Finally, the *Madres* and *Abuelas* and their supporters had the opportunity to learn the fate of their children and to find their kidnapped grandchildren. Over the years, they had reached out to activists and geneticists around the world for help in their search, notably Eric Stover,[6] then with the American Association for the Advancement of Science, who contacted the geneticists Christian Orrego Benavente, from Chile; Mary-Claire King, then at UC Berkeley; and Victor Pencheszadeh, from Argentina but living in New York. In February 1984, the *Abuelas* and other human rights organizations addressing the issue of the Disappeared convinced Sabato to ask Stover to organize a scientific team to help in the genetic identification. In June, Stover, Orrego, and King flew to Argentina with a team of renowned forensic scientists, including the forensic anthropologist Clyde Snow, to meet with judges, human rights activists, and relatives of the missing. Months later, Stover and Snow began a training program for Argentine students to begin the exhumation of individual and mass graves believed to contain the remains of the disappeared.

In Argentina, in the initial work in the aftermath of these killings and kidnappings by the junta, it was not the remains of the missing that were genotyped, because there were none, it was their progeny. And the reference samples were the parents of the missing, the grandparents of the children whose samples were genotyped in an effort to re-establish the biological familial relationships that had been disrupted so brutally by the military junta. In the pre–forensic DNA era, the geneticists and statisticians helping *Las Abuelas* developed a grandpaternity kinship score using HLA serology and

[6] Eric Stover is currently the faculty director of the Human Rights Center (HRC) at University of California, Berkeley, School of Law. Eric, my colleague Tom White, and I are editors and coauthors of *Silent Witness: DNA Evidence in Criminal Investigations and Humanitarian Disasters* (2020). Tom and I both serve on the Advisory Board of the HRC. Some of this chapter is based on the accounts in *Silent Witness*, which provides a much more comprehensive and detailed discussion of DNA testing in the Balkans and Latin America.

blood groups. Since DNA testing has been incorporated into this identification project over 25 years ago, 131 children (now young adults) have been identified, most of whom have been reunited with their biological families.

This massive genetic identification project required a database and an organizational structure. In 1987, in response to pressure from *Las Abuelas*, the Argentine Congress passed a law creating the Banco National de Datos Geneticos (BNDG) to collect blood samples from relatives and to generate genetic profiles to populate a searchable database. Under the current leadership of Mariana Herrera Pinero, a forensic geneticist, the BNDG has implemented STR genotyping, mitochondrial DNA sequencing, and Y-chromosome haplotyping to enter the profiles of over 300 families into their database. Because some of these family collections are incomplete, due to the disappearance of some family members or the death of the grandparents, limiting statistical power in the kinship analysis, the BNDG established a Forensic Anthropology Unit in 2015 to exhume and genotype skeletal remains of family members. With the increased statistical power generated by these additional profiles, the rate of identification increased significantly.

Further north, inspired by the commitment and success of human rights groups in Argentina, local organizations were formed to carry out DNA testing in the aftermath of armed conflicts in Peru, Colombia, and El Salvador. In Peru, around 20,000 people have been reported missing between 1980 and 2000. In Colombia, more than 30,000 unmarked graves have been discovered; these graves are thought to contain the remains of people killed during the civil wars and violent political repression since the mid-1960s. The BNDG has been training forensic professionals in the genetic and statistical techniques for identifying the missing in all three of these countries.

In El Salvador, BNDG and the UC Berkeley HRC, have worked closely with the *Asociacion Pro-Busqueda de Ninas y Ninos Desapareceidos* (*Pro-Busqueda*), an organization of forensic scientists, human rights activists, and relatives searching for the thousands of children separated from their families during El Salvador's civil war from 1980 to 1992. As in Argentina, these children were given up for adoption in Central America, Europe, and the United States. *Pro-Busqueda* was established by the Jesuit Father Jon de Cortina, in 1994, two years after the war's end, working with families of the missing and a Dutch human rights activist, Ralph Sprenkels. Later that year, Padre Jon, as he was known, and Sprenkels met with Reed Brody, a UN official monitoring the tenuous peace accords in El Salvador, and asked Eric Stover and his colleague Christian Orrego Benavente to help establish a DNA testing system for El Salvador. In 1995, the first DNA reunification compared the genetic profile of a boy, Juan Carlos, living in an orphanage, who had been abducted as a baby in the 1982 *Guinda de Mayo* massacres, with the profile of Maria Magdalena Ramos, who had lost her baby during that attack.

After this first successful effort, *Pro-Busqueda* continued to analyze samples, but the DNA testing program accelerated dramatically when the DNA laboratory of the California Department of Justice (DOJ) in Richmond became involved. Christian Orrego Benavente, a forensic genetics trainer at the DOJ lab, invited Stover, now the faculty director of the HRC at UC Berkeley, to talk to the lab about how DNA testing can help identify the missing using DNA databases and about the work of the ICMP in the Balkans. Stover's presentation helped establish an active collaboration between the DOJ lab and *Pro-Busqueda*, and, in 2007, the HRC forensics program, under Orrego's leadership, was awarded a grant from the US State Department to establish a DNA database and support a geneticist at *Pro-Busqueda*. The forensic geneticist Patricia Vasquez Marias began the development of databases for El Salvador and the implementation of DNA testing methods at their El Salvador lab. Patricia and Christian were married on December 12, 2013, but, tragically, Christian developed ALS (amyotrophic lateral sclerosis) and died in 2020. The ongoing work in El Salvador and other parts of Latin America is his legacy.

Unlike the DNA reunification efforts in Argentina that were supported by the government of Raoul Alfosin, which established a national genetic bank, the work of *Pro-Busqueda* was largely dependent on funding and resources provided by the HRC and the California DOJ lab.[7] The lives of thousands of children have been transformed by *Pro-Busqueda*'s work re-establishing the familial relationship that had been disrupted during the brutal civil war. One of these children who grew up in the United States, Jana Woodiwiss, currently a PhD student at the University of Georgia, found out about the DNA reunification project from the HRC, and in 2019 her biological family was located by *Pro-Busqueda*. She was able to travel to El Salvador and meet them, describing the experience in a letter to Stover, as a "life-changer."

In Ukraine in 2022, the most recent DNA technology has been brought to bear on behalf of those missing in the ongoing brutal Russian invasion.

---

[7] My colleague Sandy Calloway, her student Jessica Castrejon, and I contributed to the *Pro-Busqueda* population database by extracting DNA from blood spots sent from El Salvador and carrying out STR genotyping and statistical analyses.

Thanks to the support of a French gendarme team, the victims of Russian war crimes are being identified in mass graves and on the battlefield by a mobile DNA lab. The French team, alongside Ukrainian investigators, has been deploying a van with portable DNA genotyping instruments to gather evidence of atrocities committed on civilians. France sent the van with the DNA analysis equipment and team of forensic scientists to Kyiv in response to President Zelensky's request for the international community to investigate the massacres at Bucha.

Identifying the Missing in the United States and Canada

In the United States, the largest project on the missing is the ongoing effort to identify remains of the victims of the terrorist attacks on the World Trade Center in New York City on September 11, 2001. Over 2,753 people were killed on the ground and in the two airplanes. As a result of the explosion, only 286 intact bodies were recovered. The more than 20,000 fragments of human remains found at the site required a massive DNA testing effort, coordinated by the New York Office of the Medical Examiner. Using STR genotyping on the severely degraded remains, 1,647 individuals or approximately 60% of the victims had been identified as of March 2022. As with all of these complex missing person projects, special DNA matching software for searching reference databases was developed and implemented and, like all the other projects, the effort to identify the missing continues.

The discovery of previously unknown burial sites resulting from historical violent conflicts and human rights abuses also continues. Many of the victims in the infamous 1921 Tulsa Massacre,[8] in which around 300 people are believed to have been murdered as white mobs destroyed the large and prosperous black community known as Greenwood, have never been found. Working with ground-penetrating radar, experts identified a mass grave in the Oaklawn Cemetery in October 2020 and exhumed 19 burials. Thus far, 14 individuals have been identified for DNA extraction and analysis and descendants of the massacre victims have been asked to submit DNA samples

---

[8] Released on the hundredth anniversary of the Tulsa Massacre, the PBS documentary *The Fire and the Forgotten* explores the history of this little-known mass crime and the current attempts of the Tulsa community to come to terms with their past. Eric Stover, a coproducer of this documentary, has been in discussions with the community and the Intermountain Forensics DNA lab to address some of the concerns about privacy that have arisen.

to the Intermountain Forensic DNA lab to help identify the remains. Sadly, a group representing some of the victims' relatives, Justice for Greenwood, have expressed reluctance to submit samples, citing privacy concerns. Let's hope, in time, that these concerns can be addressed and that the mission of finally identifying at least some of the massacre victims can be achieved.

In British Columbia and Saskatchewan, the discovery of unmarked graves with the remains of indigenous children,[9] presumably students forced to attend the now defunct residential school system, has sparked a national re-examination of this dark aspect of Canadian history. These grim discoveries started around 20 years ago, when members of the *Tk'emlúps te Secwépemc* First Nation identified remains buried on the grounds of the Kamloops School in British Columbia by using ground-penetrating radar. The school, which operated from 1890 to 1997, had 500 students at its peak. The residential schools, many of which were operated by churches, banned Indigenous languages and culture, often through violence; infectious diseases, as well as sexual, physical, and emotional abuse, were widespread. Many students also died from accident, fires, and during attempts to escape, according to the National Commission on Truth and Reconciliation. The 2015 report issued by the commission after six years of research concluded that the residential school system was a "form of cultural genocide" and estimated that over 4,100 children had died or gone missing. The report demanded that the government account for all of these children. Thus far, government projects to document what happened and establish commemorative memorials are underway, but addressing this grim legacy calls for a massive program of DNA testing. Native American residential schools in the United States, like the one in Genoa, Nebraska, operated by the federal government from 1884 to 1934 have a similar dark history and will also require DNA testing to identify missing students.

The Work Continues

Even as DNA testing identifies thousands of human remains every year, the number of missing reported in armed conflicts, atrocities, and

---

[9] These unmarked graves have been found primarily by ground-penetrating radar, a method that detects anomalies in the soil, including potential human remains. Some of these discoveries may prove to be buried animals, dumps, etc., so the precise number of human remains awaits further exhumations.

natural disasters continues to grow and, sadly, their identification remains a Sisyphean task. The ICMP is currently working around the globe to identify those missing in violent conflicts, including in Iraq and Syria. They are also seeking to identify the victims and gather evidence of war crimes committed by Russian forces in the invasion of Ukraine. And previously unknown mass burial sites like those of the Tulsa Massacre and the Canadian residential schools continue to be revealed by ground penetrating radar. But the implementation of newly developed DNA technology and the concerted efforts of many local, national, and international organizations have made great progress in this essential work, continuing to roll the rock of the missing back up the hill.

PART II

RECONSTRUCTING THE ANCIENT PAST

Whereas Part I focused on how DNA analysis could help reconstruct the *recent* past, as in the crime scene, Part II describes the application of DNA analysis to reconstructing the *ancient* past, that is, making inferences about the evolution of the human species and the historical relationships among contemporary and extinct human populations. In the chapters that follow, I discuss how exploring the way genetic variation is distributed around the globe can help recreate the human migrations that gave us the modern world. The same kind of population genetics data used to estimate the random match probability (RMP, that is, the probability of a random individual also matching the crime scene evidence profile, a calculation based on the frequency of a specific genotype in different populations) can reveal how modern humans migrated out of Africa and populated the rest of the world, and can shed light on our concepts of race, ancestry, and ethnic identity.

The comparative analysis of DNA sequences, the linear order of bases in the genome, makes it possible to estimate the sequence *in time* of critical events in the history of populations or species. These histories can be inferred from the extent and pattern of differences between the sequences of different individuals within the same species or from different species. Until recently, these analyses have been limited to DNA from contemporary species. Within the last 30 years, the ability to extract, amplify, and analyze DNA from ancient specimens, including archaic human subspecies like the Neanderthals, has provided dramatic new insights into human history. This research on ancient DNA is discussed in Chapter 10 and the chapters that follow in the context of the pioneering role of Allan Wilson and Svante Paabo. The development in the mid-1980s of the PCR method for amplifying specific DNA sequences from human genomic DNA and the more recent development, dating from 2005, of next generation sequencing (NGS) were the technological breakthroughs that made these kinds of genetic analyses pos-sible. This field is moving so rapidly that new observations and developments will inevitably provide more detail, nuance, and complexity to the picture discussed here.[1]

---

[1] An excellent review of DNA analysis and human evolution as of 2018 is David Reich, *Who We Are and How We Got Here: Ancient DNA and the New Science of the Human Past* (2018).

9

Allan Wilson, Molecular Evolution, and the Out of Africa Hypothesis

Humans are found in every corner of this planet. How did we get here (or there) and where did we all come from? Based on the fossil record, there is universal agreement that our ancestors came from Africa, but until recently, there were two competing narratives for how modern humans populated the rest of the world. One theory, known as the multiregional model, was that early hominin forms, such as *Homo erectus*, migrated from Africa millions of years ago to Europe and Asia and that these archaic subspecies evolved more or less independently into anatomically modern humans, *Homo sapiens*. The alternative theory, the replacement model, was that modern humans emerged relatively recently from Africa, *replacing* the archaic subspecies, such as Neanderthals, that had populated Europe and Asia, eventually spreading to the Pacific and the Americas. In a sense, these competing narratives both assumed an African origin of modern humans but differed in *when* the migration started and *where Homo sapiens* evolved.

The relationships among a set of DNA sequences from different populations or species can be represented as a tree-like structure with branches and a root, a so-called phylogenetic tree. It seems only fitting that we can visualize evolution using the image of those majestic living things, which, by virtue of their ability to convert water, sun, and carbon dioxide from the air to sugar while releasing oxygen, have made the diversity of life on earth possible. Treelike diagrams have been used for centuries to represent the evolutionary relationships of different living species, as well as the hominid fossil record and, even, of nonbiological entities, such as languages. DNA sequences, however, offer significant advantages for the construction of phylogenetic trees; precise objective quantitative analysis of relationships can be estimated by simply counting the number and pattern of base changes. One of the most important insights about human history inferred from DNA sequences in a phylogenetic tree is the by now well-known Out of Africa hypothesis proposed in 1987 by Allan Wilson and his graduate students

Rebecca Cann and Mark Stoneking[1] at UC Berkeley (Cann, Stoneking, and Wilson, 1987). (I suspect Wilson enjoyed the resonance of the name of this hypothesis with the classic Isak Dinesen novel of the same name.) The idea, based on DNA analysis of mitochondrial DNA—the DNA present in the cellular organelle the mitochondrion—that all anatomically modern humans (*Homo sapiens*) are derived from a group of Africans that migrated out of Africa within the last 75,000 years or so (estimates vary) and colonized the globe, was consistent with some interpretations of the fossil record but not with others.[2] The Out of Africa model was based on the pattern of DNA sequence diversity in the mitochondrial DNA phylogenetic tree (Figure 9.1). Cann, Stoneking, and Wilson noted that there was more diversity (more different DNA sequences) within the African branch of the tree than in all the other branches representing the rest of the world. The diversity outside Africa therefore, they concluded, represented only a *subset* of the diversity within Africa. Their interpretation of this pattern was that all modern humans are the descendants of a small subpopulation of Africans that left Africa recently (about 75,000 years ago) and proceeded over time to populate the rest of the world. The Out of Africa hypothesis caused a considerable stir in the evolutionary community, consternation among the multiregional model proponents, and intense interest in the general public's transformed sense of human origins. *We are all Africans! And relatively recent immigrants.*

The Molecular Clock

Wilson, a charismatic New Zealander with long silver hair, was a highly creative and innovative evolutionary biologist who pioneered the analysis of, initially, proteins and then, later, DNA to make inferences about human history. With his colleague Vincent Sarich, he demonstrated that the concept of a "molecular clock," based on differences in the sequence of DNA and/ or proteins of living species, could be used to estimate the time of species divergence from a common ancestor. (The idea of a "molecular clock" had

---

[1] Mark Stoneking later came to my lab as a postdoctoral fellow and developed the first mitochondrial PCR-based DNA typing system, which was used initially to identify the remains of a missing young girl, discussed in Chapter 8.

[2] All modern humans *outside of Africa* are derived from this migration. Of course, modern Africans and the members of the African Diaspora, including the 18th- and 19th-century slave trade, are mainly descended from groups who remained in Africa (*remainers*). A small number are the descendants of back-migrations into Africa.

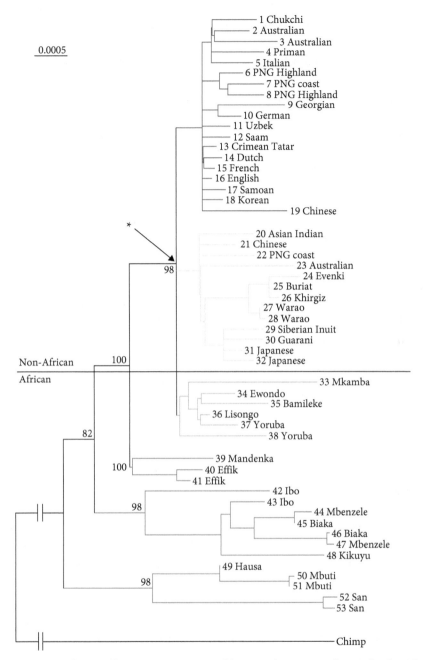

Figure 9.1 This tree from a more recent publication (Ingram et al., 2000) is based on the analyses of 53 different individuals and confirms and extends the initial mitochondrial DNA analysis from the Wilson lab. Unlike the earlier publications, which focused on the so-called hypervariable region of the mitochondrial genome, this study analyzed the *complete* mitochondrial genome, but excludes the hypervariable region. The high mutation rate in the hypervariable region makes it difficult to assign time estimates. Individuals of African descent are found below the dashed line and non-Africans above. The node marked with an asterisk refers to the Most Recent Common Ancestor of the youngest clade containing both.

been proposed initially by Emile Zuckerkandl and Linus Pauling in 1962.) The basic idea underlying the "DNA clock" was that one could compare sequences of the same gene from two closely related species and, assuming that DNA sequences accumulated differences by random mutation at some specified constant rate over time, estimate when the two species diverged from a common ancestor. These analyses showed that the split between the evolutionary line that led to humans and the one that led to chimpanzees, our closest living relative, was much more recent (about 5 million years ago) and that the divergences of these two lines from that of the gorilla lineage was also much more recent (about 7 million years) than the divergence times that the evolutionary biology community had previously assumed (around 20 million years), based on the fossil record.

Wilson was one of my scientific heroes. He was a bold and brilliant innovator and a charismatic and inspiring mentor. The Wilson lab was a lively and stimulating place, and the regard and affection that his students felt for him and for one another reminded me of my days as a postdoc at Stanford in Hugh McDevitt's immunogenetics lab. Some labs, usually really good ones, develop a family structure with succeeding generations of students and postdocs regarding one another as relatives (big brothers, crazy aunts, rival siblings—the full familial spectrum). Several former members of his lab at UC Berkeley, notably Norm Arnheim, Russ Higuchi, Alice Wang, and Tom White, played crucial roles in the development of PCR technology at Cetus, and an ongoing and stimulating interaction developed between Allan and our Cetus PCR group.

Two of my brightest postdoctoral fellows, Ulf Gyllensten and Steve Mack, came from the Wilson lab, bringing with them the concepts and tools of molecular evolution and applying them to studies of the HLA genes, discussed briefly in Part I. These are the genes involved in the immune system that require precise matching between donor and recipient for bone marrow transplantation. They are the most polymorphic genes in the human genome, and one of them (the HLA-DQA1 gene) was used in the first DNA case (*Pennsylvania vs. Pestinikas*, 1986) described in Chapter 1. I found our studies on the evolution of the HLA genes stimulating as well as great fun, and thinking about the complexity of the immune system from an evolutionary perspective was rewarding at many different levels. As the evolutionary biologist and fruit-fly geneticist Theodosius Dobhzhansky famously said, "Nothing in biology makes sense except in the light of evolution."

Ulf and Steve brought their expertise in constructing and interpreting phylogenetic trees, a crucial analytic tool, to the lab. The structure of these phylogenetic trees reflects the pattern of differences in the individual DNA sequences, but there are many different ways to construct such trees. Typically, the "genetic distance" (the difference between the DNA sequences) between two sequences is calculated for all possible sequence pairs and then one of many possible algorithms is used to construct a tree connecting all the sequences. A sequence for an "outgroup"—a group that is evolutionarily distant from the sequences under analysis—is used to "root" the tree. In the mitochondrial DNA tree (Figure 9.1), the chimpanzee mitochondrial DNA sequence serves as the outgroup used to root the tree for the human sequences. With this structure, we can distinguish *ancestral* from *derived* sequences, DNA sequences that have evolved from the ancestral sequence by mutation, based on the position of the sequence within the tree, and we can determine the number and nature of the DNA sequence changes that have occurred along each branch.

Cann and Wilson analyzed mitochondrial (mt) DNA—the 16,000-base-pair circular genome within the mitochondria, the organelles that produce chemical energy (ATP) within each cell—from 147 individuals from a variety of global populations. MtDNA is, as noted in Part I, present in hundreds or thousands of genome copies per cell and is inherited through the maternal lineage. The fertilized egg contains mitochondrial DNA from only the mother.[3] Siblings inherit the same mtDNA from the mother, but only females will transmit this sequence to the succeeding generations. The pattern of the mtDNA tree (Figure 9.1) is that all of the branches from non-African mtDNA sequences are a clustered subset of all of the sequences and the root of the tree is African.[4]

Working in the pre-PCR era, Wilson and his colleagues had taken the approach of physically isolating mtDNA from the chromosomal DNA and, rather than directly determining the sequence, indirectly investigated the sequence differences between individuals by RFLP (restriction fragment length polymorphism) analysis (see Chapter 1). With PCR, one could simply amplify a segment of the mtDNA from each of the individuals, generating relatively pure mtDNA, thereby eliminating the need for physical isolation from chromosomal DNA. In 1991, Linda Vigilant, Mark Stoneking, and

[3] In a few very rare cases, a paternal mtDNA contribution has been detected.
[4] All of the human sequences are equidistant from the chimpanzee outgroup sequence.

others in the Wilson lab could then directly determine the sequence of the amplified mtDNA from different individuals in many different populations and construct a phylogenetic tree from all of these sequences (Vigilant et al., 1991). This direct comparison of mtDNA sequences provided much better statistical evidence for the Out of Africa hypothesis and the inferred timeline than did the original analysis based on RFLP.

The structure of the tree in which all non-African populations represent a clustered subset of all the sequences present in Africa was interpreted to mean that a portion of the modern human groups living in Africa had migrated out of Africa, colonized the rest of the world, and that all modern humans were the descendants of this migration, along with those who remained in Africa. The observation from the mtDNA tree that there was more genetic diversity within Africa than in the rest of the world because this Out of Africa migration represented a genetic bottleneck (a sharp reduction in the population size leading to a reduction in genetic diversity) has been supported by all subsequent genetic analyses. Virtually all genes analyzed in different human populations show that there is more sequence variation, or genetic diversity, within Africa than outside Africa. These population bottlenecks are a common feature of human migrations. The human colonization of previously unpopulated areas, such as Polynesia, Australia, or the Americas by small founding populations, also known as the "founder effect," results in reduced genetic diversity. Also, severe reductions in population size by disease or war can reduce the genetic diversity of the surviving population.

When the Wilson Out of Africa hypothesis, based on the mtDNA phylogenetic tree, was first proposed, a lively and occasionally rancorous debate ensued between proponents of the replacement and multiregional models. The time frame estimated for the Out of Africa migration (around 75,000 years ago) was completely inconsistent with the notion that modern humans evolved in different parts of the world over much longer time periods. I recall hearing polemical lectures by Milford Wolpoff, a paleoanthropologist who was a fierce defender of the multiregional model, that would have been better suited to a political rally than a scientific conference. I think it's safe to say now that the replacement model, with a few important and surprising refinements (such as the interbreeding between Neanderthals and modern humans to be discussed in Chapter 10), has prevailed and is consistent with both the genetic data and the most recent interpretations of the fossil record.

All phylogenetic trees have a "root," a node where all the branches converge or "coalesce"; this mtDNA sequence from which all contemporary mtDNA sequences are derived became known as "Mitochondrial Eve" or "African Eve." A friend and graduate of the Wilson lab recently told me that Wilson himself found the "Eve" appellation unfortunate; the apparent origin of "Eve" was a tabloid headline, so perhaps a nameless journalist should get the credit—or the blame. Whatever the source, the African "Eve" quickly captured the popular imagination; a *Newsweek* magazine cover at the time ("The Search for Adam and Eve") featured an attractive black couple, looking very much like the then popular R&B duo Peaches and Herb, cavorting in an Eden-like garden.

It is in the nature of an mtDNA tree that all contemporary branches would ultimately coalesce to or emanate from a single ancestral mtDNA sequence. Since mtDNA is transmitted only though a female child, any mtDNA lineage from a woman with no daughters will go extinct; therefore, that there is a Mitochondrial Eve, a single ancestral root sequence, is the result *expected* from *any* mtDNA tree and is not really "news." *Further, this pattern for the mtDNA phylogenetic tree says nothing about the evolution of our chromosomal DNA.* Nonetheless, the misconceived but provocative idea that we all descended from this single African woman was how some people—including some who should have known better—interpreted the Mitochondrial Eve and Out of Africa narrative.

The real news was not that all the mitochondrial DNA sequences coalesced into a single "root" but instead that this root is African and that a rough estimate could be imposed on the date of the Out of Africa event. The chronological distance to a common ancestor revealed that all modern humans are descended from a group of Africans who either remained in that continent or migrated from Africa around 75,000 years ago, though current estimates vary from around 50,000 to 100,000 years ago. The coalescent time reveals the time when the common mtDNA ancestor lived and represents an upper bound for the time for the Out of Africa migration. This date is much more recent than the dates of the oldest fossil remains of anatomically modern humans, estimated at around 200,000 years ago. A 195,000-year-old fossil from the Omo12 site in Ethiopia shows a skull morphology associated with modern humans.

Coalescent times can be estimated by the pattern of sequence differences in this and in other phylogenetic trees using the assumption of a constant "molecular clock" that relates the sequence differences to a time scale. The "clock" can be calibrated by considering the number of sequence differences between, for example, the same gene from Old World monkeys and New World monkeys, assuming a species separation time from their common ancestor of, say, 40 million years, based on radiocarbon dating of the fossil record. The molecular clock model makes two critical assumptions: (1) that the ancestral species had an ancestral gene sequence that diverged in the lines leading to the two contemporary species and (2) that the sequence differences being used to estimate time are not themselves subject to natural selection. Natural selection could make sequence divergence faster (diversifying selection) or slower (purifying selection) than the rate of neutral evolution, which reflects observed mutation rates and population size, which determines how likely a new randomly generated mutation will persist in the population. The concept of "neutral" evolution refers to DNA sequence differences that are nonadaptive (i.e., confer no advantage) and are not deleterious and therefore not subject to natural selection. So molecular clocks and the time estimates derived from them assume neutral evolution. This assumption is absolutely critical to the interpretation of DNA sequence data using the molecular clock concept. Ignoring this essential assumption, as we'll see below, only invites confusion and trouble.

Neutral Evolution

The whole concept of "neutral evolution," sometimes known as "non-Darwinian" evolution because it proposes that some evolutionary changes are "nonadaptive," serves as a useful corrective to the idea that all evolutionary change necessarily increases adaptation and fitness. Its acknowledgment of the role of random change in evolution has a philosophical resonance. To the worldview that "everything happens for a purpose," neutral evolution theory responds, "No, some things just happen." A number of population geneticists, notably Motoo Kimura and Tomoko Ohta, have developed elegant and elaborate mathematical models for the neutral theory of molecular evolution. A common strategy for demonstrating the operation of natural selection on DNA sequence data is to compare the observed data with the mathematical predictions of a neutral evolution model. If the observed data differ significantly from the neutral prediction, we can infer that natural selection is at work. Mathematics aside, the notion that some nonadaptive genetic variants increase in the population solely by chance can provide some

perspective when considering a society in which some successes do not reflect merit, skill, talent, or privilege, but simply the operation of chance.

Phylogenetic trees and molecular clocks assume an evolutionary model
in which DNA sequences diverge by the gradual accumulation of individual base substitutions. DNA sequences, however, can also evolve by recombination, the exchange of DNA segments between chromosomes that
takes place during meiosis in the formation of gametes. Another kind of
genetic exchange is the much rarer so-called horizontal transfer in which
DNA segments can be transferred from one individual organism to another or even between species. These gene transfers are termed "horizontal"
to distinguish them from the well-known "vertical gene transfer" involved
in parent-to-child transmission. Such horizontal gene transfers are much
more common in the bacterial world than they are in the human. A recent
book on evolutionary genetics, *The Tangled Tree* (2018) by David Quammen
discusses how well-documented evidence of gene transfer between species
in different branches of the phylogenetic tree may call into question not only
the treelike representation of species evolution but the basic Darwinian concept in which species evolve over time, separating into different lineages
based on the gradual accumulation of mutations. Horizontal gene transfer
is, certainly, a fascinating natural phenomenon and Quammen's question a
provocative one. But most of the DNA in the genome of most organisms is
not involved in these events. Gene exchanges between chromosomes of the
same or of different individuals notwithstanding, the branching pattern of
the phylogenetic tree based on the simplifying assumption of the random accumulation of mutations and the time estimates based on a molecular clock
can provide extremely valuable historical insights.

If the molecular clock requires comparing DNA sequences that are not
subject to natural selection, how then does one distinguish neutral changes
from those that reflect selective pressure? One way to is to consider DNA
sequence changes in a gene that do not change the structure of the protein
encoded by that gene. The nature of the genetic code and the structure of
genes provide two different strategies for identifying sequence changes that
might be "neutral."

In order to understand the molecular clock, it's essential to understand the
genetic code. And what could be more intellectually exciting than breaking
a code? Whether it's Alan Turing cracking the Enigma code or Jean-Francois
Champollion deciphering the Egyptian hieroglyphics in the Rosetta Stone,
code-breaking serves as a romantic metaphor for the scientific enterprise

itself. The genetic code refers to the way information in the base sequence in DNA is expressed as protein structure and is the program used by all living things on earth. The scientific articles that revealed the genetic code were published while I was an undergraduate; reading those papers is what inspired me to become a geneticist. One of my favorite novels, *The Gold Bug Variations* by Richard Powers, creates a brilliant and moving narrative out of the search for this fundamental code.

What follows here may be tough going for the nonscientist but the fundamental nature of this code, in my view, warrants the effort. Here goes. The information in DNA is expressed when the sequence of bases in a gene is "translated" into the sequence of amino acids in proteins, via an intermediary, "messenger RNA." Amino acids are the building blocks of proteins, the molecules that provide many of the structural elements and carry out most of the functions within the cell. There are 20 different amino acids. The genetic code specifies how that translation is carried out. In 1961, Francis Crick, Sydney Brenner, Leslie Barnett, and R. J. Watts-Tobin, demonstrated in a series of conceptually elegant but remarkably simple experiments with bacterial viruses that the information in the DNA sequence is "read" with a nonoverlapping triplet code. A series of three bases, now known as a "codon," codes for an amino acid. With 4 different bases (A, C, T, G), there are 64 (4^3) possible triplet codons, but there are only 20 different amino acids. Therefore, some amino acids are encoded by several different triplet codons.[5] This code, in which multiple triplet codons can correspond to a single amino acid (for example, the codons GGA, GGC, GGG, and GGT all encode the same amino acid, glycine), is known in information theory as "degenerate." (No value judgment implied.) This property of the genetic code makes it possible to distinguish between DNA sequence differences that might be under selection from neutral changes. DNA sequence substitutions that change the amino acid sequence of a protein are known as "nonsynonymous" (e.g., GGA, which codes for glycine, to CGA, which codes for arginine) while those that don't are termed "synonymous" or "silent" (e.g., GGA to GGC, both of which code for glycine). Since proteins carry out many of the functions in the cell and individual organism, changes in protein structure might well be subject to natural selection while silent or synonymous DNA sequence changes are, in general, assumed to be neutral.

---

[5] Three of these 64 different codons (TAG, TAA, TGA) tell the translational machinery to stop, and one (ATG) tells it to start. In 1964, Marshall Nirenberg and Phil Leder, in a series of sophisticated experiments, identified which triplet codons specify which amino acid.

Another source of potentially neutral DNA sequence changes are sequences within the gene that do not code for protein. In 1977, a postdoctoral fellow in Richard Flavell's lab at the University of Amsterdam discovered that the rabbit beta-globin gene was interrupted. We now know that virtually all mammalian genes are "split"; portions of the gene that contain triplet codons specifying the sequence of amino acids in the translated protein are separated by DNA sequences that do not. These intervening sequences (initially called, rather unimaginatively, "intervening sequences") are now known as "introns," in contrast to the gene segments known as "exons," that are expressed as protein sequences.[6] Since introns do not code for protein, sequence differences in the introns[7] as well as synonymous sequence changes in the exons can be considered to be neutral and used for molecular clock calculations. Most, although not all, of the DNA sequence difference outside genes (the vast majority of the 3 billion bases of the human genome) might also be considered to be neutral, although we now know that many of these sequences regulate the expression of neighboring genes.

Ancestral Polymorphism and
Origin of the Human Species

The type of DNA sequence analysis embodied in phylogenetic trees of different species can be applied to the evolution of a specific gene by examining DNA sequences from closely related species. The HLA genes that I had been studying for many years are by far the most polymorphic genes in the human genome, and I had often wondered about how this remarkable allelic diversity might have evolved. I was fortunate that the very knowledgeable evolutionary biologist Ulf Gyllensten chose to come to my lab for a postdoctoral fellowship in the mid-1980s after having studied with Allan Wilson. We decided to sequence some of the HLA genes from different individuals from several different nonhuman primate species, such as chimpanzees, gorillas,

---

[6] Following the transcription of the gene sequence (exons + introns) into RNA, the intron sequences are "spliced out" of the messenger RNA, which is then translated into protein. The 1993 Nobel Prize in Medicine and Physiology was awarded to Phillip Sharp and Richard Roberts for the discovery of split genes and RNA splicing. (Roberts, as noted in Chapter 2, was as an expert witness for the defense in *People vs. Castro*, while the postdoctoral fellow who showed that the beta-globin gene was interrupted with nonprotein coding introns was none other than Alec Jeffreys, now Sir Alec, discussed in Chapter 1 as the pioneer of DNA fingerprinting.)

[7] Specific exemptions are the sequences at the exon/intron junction involved in RNA splicing that may be subject to natural selection.

bonobos, etc. (The H in HLA stands for "human" so the homologous genes in other species have different names, but for the sake of simplicity, I refer to the homologous genes in other species by their human name as well.) Initially, we focused on HLA-DQA1, the same gene we had used in the first forensic DNA case (*Pennsylvania vs. Pestinikas*, 1986).

Our work to understand the evolutionary origins of the diversity of human HLA-DQA1 alleles provided one of the many connections between our research on evolution and our forensic work. As it turned out, we could use the same PCR system we employed for amplifying the human HLA-DQA1 gene to amplify the homologous gene for these other primate species. These evolutionary studies established population genetic data that served as the foundation of the forensic application of this gene, and of the commercial forensic HLA-DQalpha test kit we launched in 1991. For several years, this was the only PCR-based genetic test available to the forensic community, and it was very useful to know how the alleles were distributed in different human populations and how they had evolved.

When we analyzed the sequences of HLA-DQA1 as well as other HLA genes from various primate DNA samples, we found that they were also highly polymorphic in all of the primate species we studied. This was not unexpected. One of the arguments for explaining the large number of HLA alleles in human populations was that the individuals and populations with a wide variety of alleles could mount a more effective and diverse immune response to infectious disease pathogens and that, therefore, natural selection favored HLA allele diversity. If HLA diversity was good for humans, then it was reasonable to expect that the same might be true for other species.

One of the unexpected findings from this phylogenetic analysis was that there were some human HLA-DQA1 sequences that were more closely related to specific chimpanzee or gorilla sequences than they were to other DQA1 human sequences. This pattern was observed by a few other scientists and one of them, Jan Klein, an imposing, authoritative, and argumentative figure in the fields of immunogenetics and evolution, termed it "transspecies evolution," which I found confusing, even misleading. I preferred "ancestral polymorphism," which has prevailed, as it captures the notion that the ancestral species that had given rise to contemporary primate species carried these many different HLA alleles and had transmitted them relatively unchanged to the contemporary primates. Once again, if natural selection had favored diversity in modern primate species, why not in our ancestors as well? To illustrate this concept, consider an ancestral species that contains

HLA variants A, B, C, D, and E and transmits some or all of these variants with minimal changes to chimpanzees, gorillas, and humans. A human version of the A variant will be more like the chimpanzee or gorilla A variant than it would be like the human B, C, D, or E variant.

What seemed like a rather arcane question concerning the number of HLA alleles transmitted from the ancestral species to modern humans turned into a highly contentious dispute about the origins of the human species between Ulf and me on the one hand and on the other Jan Klein and the population geneticist Francisco Ayala. Ayala and Wilson were long-standing rivals in the molecular-evolution community and, in his 1995 *Science* article "The Myth of Eve," Ayala attacked with great relish an overly simplistic ("straw-woman") version of the Out of Africa hypothesis. The abstract for that paper starts out, "It has been proposed that modern humans descended from a single woman, the mitochondrial Eve, who lived in Africa 100–200,00 years ago" and concludes that "the mitochondrial Eve hypothesis emanates from a confusion between gene genealogies and individual genealogies."

The article proceeds to an interesting discussion about the highly polymorphic HLA genes, the age of the alleles, and the size of the early human population. In the emergence of a new species, the approximate size of the founding population can be roughly estimated based on the number of alleles transmitted to the new species by the ancestral species. (The population genetics equations use a concept known as the "effective population size," an estimate of the number of reproducing organisms in the larger population at any one time.) The larger the number of such "ancestral" alleles, the larger the founding population would have to have been to accommodate them. If the human species originated with Adam and Eve in the Garden of Eden, the *maximum* number of founding human alleles would have been four. Or just two, if one chose to incorporate Eve's emergence from Adam's rib in the calculation. For non-Genesis-based accounts, determining the number of human alleles that were ancestral or predated the divergence of humans from chimpanzees 5 million years ago could help provide rough estimates of the founding population size.

Ayala focused on the HLA-DRB1 gene and argued that a large number of HLA-DRB1 alleles diverged from one other >6 million years ago, based on the sequence differences between them. Since this time estimate for allele divergence (>6 million years) was greater than the estimate for chimpanzee and human divergence (5 million years ago), Ayala inferred that this large number of HLA-DRB1 alleles had been transmitted from the ancestral species to humans. Based on these assumptions, he estimated that the effective population size was very large (>100,000). This estimate, based on his analysis of the HLA-DRB1 alleles, was much larger than the population sizes estimated from analysis of mtDNA diversity by the Wilson group and from the analyses of other gene sequences. These population size estimates were <10,000. While I was delighted to see this focus on my favorite set of genes, the discrepancy between the estimate based on HLA genes and other genes was, as discrepancies always are, worrisome. So, Ulf and I set ourselves the task of trying to account for these substantial differences. I suppose we were also being somewhat defensive of the Wilson legacy.

The age estimate reflects the time at which the alleles diverged from one other, based on the sequence differences between the HLA alleles and the structure of the phylogenetic tree. This time estimate, in terms of the molecular clock, looks "backward" to the "coalescence" time, the time to the closest common ancestor. The more divergent the sequences, the greater the time to the most recent common ancestral sequence. Ulf and I considered whether the source of the discrepancy was that Ayala had applied the molecular clock to DNA sequences that had been under natural selection rather than restricting the analysis to evolutionary neutral sequences. Ayala's phylogenetic tree used the highly polymorphic exon 2 sequences of HLA-DRB1; these sequences are known to be subject to strong natural selection.[8]

We wondered whether Ayala had overestimated the number of ancestral HLA alleles based on considering sequence differences that had been subject to diversifying selection. So, with Ulf's student at the University of Uppsala, Tomas Bergstrom, we decided to make a phylogenetic tree from HLA-DRB1 intron sequences and apply the molecular clock to these presumably neutral sequence changes. This analysis revealed that there were relatively few ancestral HLA-DRB1 sequences that predated the evolutionary split between humans and chimpanzees (Bergstrom et al., 1998). The population size we inferred, based on these neutral intron sequence differences, was much smaller (<10,000) and was, in fact, consistent with the other genetic data.

---

[8] The role of the protein encoded by the HLA-DRB1 gene is to bind and present foreign antigens, like a fragment of a viral protein, to the immune system, initiating the immune response to that specific viral protein. The part of the HLA-DRB1 molecule that binds foreign antigen is encoded by exon 2; the extensive allelic diversity of exon 2 is thought to reflect diversifying natural selection to maximize the number of different infectious disease pathogens recognized by the immune system of individuals, populations, and species. The operation of natural selection on these exon 2 sequences was revealed in an elegant statistical analysis by the population geneticist Masatoshi Nei, who showed that the ratio of "nonsynonymous" DNA changes to "synonymous" changes was much greater than expected for a neutral model.

All of the phylogenetic analyses, like the phylogenetic tree with the mtDNA (Figure 9.1) or the HLA gene genealogy discussed thus far use sequences derived from DNA samples from living individuals and extant species. In the structure of the tree, one could, in principle, *infer* the sequence of an ancestral gene taken from an ancient or even extinct species. The development of PCR, however, made the *direct* genetic analysis of ancient samples possible. It was Wilson who, at a dinner with my Cetus colleagues Norm Arnheim and Tom White, first suggested that PCR could be used to analyze ancient DNA samples for the purposes of evolutionary inference.[9] Even in the pre-PCR days, Wilson's postdoctoral fellow Russell Higuchi had extracted and analyzed DNA from a museum skin specimen of a quagga, a zebra-like extinct species. But it was PCR and, two decades later, the development of next generation sequencing that transformed the study of old bones and museum specimens from a curiosity into the burgeoning field of ancient DNA analysis. This research has revealed previously unknown archaic species and unexpected interactions between *Homo sapiens* and these archaic species (e.g., Neanderthals).

In 1991, Allan Wilson was diagnosed with leukemia. A bone marrow transplant was considered, provided an HLA-matched donor could be found. In a very personal connection between our use of PCR in forensics and clinical medicine and the medical treatment of this evolutionary biology pioneer, our lab carried out PCR-based HLA typing of hairs sent to us from two of Allan's family members in New Zealand. It turned out that one of his brothers was an HLA-identical match, and Gary Wilson flew from Auckland to Seattle, where Allan was being treated. Allan received a bone marrow transplant at the Fred Hutchinson Cancer Research Center in Seattle but, sadly, he died of a post-transplant infection on July 21, 1991, at the age of 56. His impact on evolutionary biology was immense, and his legacy, both his ideas and the scientists he trained, can be seen in the vibrant field of molecular evolution he helped create, and particularly in ancient DNA studies, one of the most exciting and productive fields in human biology and history.

[9] At that dinner, four months before the foundational article on PCR was published in 1985, White suggested that our new method for DNA amplification, PCR, might be able to help identify the burned remains of victims found at the scene of a crime widely discussed in the newspapers. Wilson responded that, if so, PCR could also be used to analyze the remains of extinct species in museum collections. Intrigued with the potential of PCR for molecular evolution, White took a one-year sabbatical leave from Cetus and attended the Wilson lab meetings at UC Berkeley and, with Professor John Taylor, studied the evolution of fungi, organisms that lack a fossil record, in order to establish taxonomic relationships based on DNA, publishing a seminal paper on DNA barcoding and phylogenetic analysis (White et al., 1990).

10

DNA of the Dead

Sequencing Archaic Species and Ancient Remains

In 1984, Russell Higuchi,[1] a postdoctoral fellow in the Wilson lab and, later, a close friend and colleague at Cetus and Roche Molecular Systems, reported generating short mitochondrial DNA (mtDNA) sequences from a museum specimen of the quagga, a zebra-like species (*Equus quagga*) that became extinct in 1883 (Higuchi et al., 1984). Based on comparing the sequence with that of a mountain zebra, he and his colleagues concluded that the two species had a common ancestor 3–4 million years ago, consistent with the fossil evidence. This landmark paper analyzing the DNA of an extinct species was reviewed in *Nature* by Alec Jeffreys, the DNA fingerprinting pioneer, under the title "Raising the Dead and Buried." Russ's job seminar at Cetus in late October 1986 was titled "DNA of the Dead," accompanied by appropriately spooky Halloween graphics and is the inspiration for this chapter's title.

The development of PCR was first reported a year after the quagga paper and, just as Wilson had envisioned, made it possible to unlock the secrets of ancient DNA on a scale that had been previously unimaginable. After Russ joined Cetus, we transferred PCR technology to the Wilson lab and, in 1989, Russ, Wilson, and a Swedish postdoctoral fellow in the Wilson lab, Svante Paabo, published a paper, "Ancient DNA and Polymerase Chain Reaction: The Emerging Field of Molecular Archaeology" (Paabo, Higuchi, and Wilson, 1989). In the abstract, they proclaimed that new "molecular biological techniques have enabled us to retrieve and study ancient DNA

---

[1] While at Cetus, Russ developed a quantitative method for polymerase chain reaction (PCR) amplification. This novel modification of the basic PCR method, known as qPCR or real-time PCR, was used for quantitative infectious disease diagnostic tests, including the HIV viral load test developed by John Sninsky and Shirley Kwok. The HIV viral load test was, in turn, critical for the evaluation and development of new anti-HIV drugs. qPCR has also been used in the analysis of waste water to assess the level of the COVID-19 virus within a community. This application is known as environmental PCR or ePCR. Russ also contributed to the development of our HLA-DQalpha forensic test and to the first DNA analysis of a single hair. In 2019, Russ received the Award for Excellence in Molecular Diagnostics from the Association of Molecular Pathology. Alec Jeffreys received this same award in 2003, as did I in 2000.

molecules and thus to catch evolution red-handed." In 2022, Svante Paabo, the director of the Max Planck Institute for Evolutionary Anthropology at Leipzig, was awarded the Nobel Prize in Medicine or Physiology based on his contributions to the ancient DNA field and, in particular, to the sequencing of the Neanderthal genome.

In the early 1990s, a flurry of papers was published reporting the extraction and the PCR amplification of DNA from ancient animal and plant samples. In the Wilson lab, Svante Paabo had proposed analyzing the DNA of Egyptian mummies, the first attempt at ancient human DNA analysis. Amid concerns about the theoretical chemical stability of DNA and about the potential for contamination of these ancient specimens with "modern" DNA, these reports elicited both excitement and skepticism. In 1993, the Swedish biochemist Tomas Lindahl published a review article in *Nature* suggesting that the "occasional dramatic success" in this area should be viewed with skepticism. He argued that the "widely publicized report of the recovery of 17–20 million year old DNA from an ancient plant leaf exhibits serious deficiencies," noting that this reported observation "was incompatible with the chemical structure of DNA" (Lindahl, 1993, p. 709). With the potential of extracting and amplifying DNA from ancient specimens, some scientists proposed resurrecting some extinct species like the wooly mammoth or even dinosaurs. There is a fine line between visionary and crank . . . and some of the early enthusiasts of ancient DNA research may have crossed it.

During this heady time when ancient DNA research promised a new and exciting frontier, PCR amplification of DNA from ancient insects embedded in amber, a presumably protective condition for the stability of DNA, was also reported. The notion that such an insect might conceivably contain blood obtained by biting a dinosaur was suggested by the UC Berkeley entomologist George Poinar, whose imagination was rather less constrained than his more biochemically oriented colleagues.[2] This conceit, of course, became, with a few elaborations, the basis for Michael Crichton's *Jurassic Park* in 1993.[3] Evidence to support this suggestion, if not the film's narrative, was reported in *Nature Communications* in 2017. David Grimaldi, an entomologist

---

[2] Poinar visited the lab in Cetus with a piece of amber with an embedded insect, eliciting a lively discussion with Russ and me.

[3] When the movie was released, I was interviewed for a local TV news program. I described how PCR worked and, in response to a question, explained that recreating dinosaurs from DNA in amber was pure science fiction . . . wildly imaginative but without any scientific basis. When the interview aired, my explanation of PCR survived but my skepticism was deleted. Nobody likes a party-pooper.

at the American Museum of Natural History, identified a piece of amber from northern Myanmar in a private collection with a 99-million-year-old tick grasping a dinosaur feather. He and his colleagues speculated that the tick's host was a fledgling nonavian dinosaur, no bigger than a hummingbird. Recovering dinosaur DNA, however, remains the realm of fiction.

Notwithstanding potential dinosaur DNA in amber-embedded insects, human fossil remains were arguably the most exciting potential source of ancient DNA. If recovering and amplifying DNA from a million-year-old specimen was highly unlikely, DNA that was thousands or even hundreds of thousands of year old might still allow the possibility of PCR amplification. The major problem with these early heroic attempts to amplify and analyze ancient DNA was due to the handling by anthropologists, museum staff, and scientists to which these bones had been subjected. Contamination with contemporary human DNA on the surface of the bone gave rise to the very high probability, given the minute amount and degraded nature of whatever ancient DNA could be recovered from the sample, that any DNA successfully amplified would be derived from the handlers. Svante Paabo's laboratory, the acknowledged leader in the field of ancient DNA research, was clearly the most committed and rigorous group to attempt to amplify ancient human DNA and to minimize contamination during this period. (Henri Poinar, the son of the insect-in-amber-champion George Poinar, was an important member of this lab.) After many attempts at amplifying mtDNA from Neanderthal bones, most of which resulted in obtaining amplified DNA from human handlers, they were finally able to successfully amplify and sequence some short mtDNA fragments. Because of the significant divergence from any modern human mtDNA, these mtDNA sequences were almost certainly derived from the ancient Neanderthals who were the source of these museum specimens (Green et al., 2008). This was universally and deservedly recognized as a major tour de force.

The results of these studies were far more meaningful than simply a dramatic technical achievement. The two different mtDNA sequences obtained from two different Neanderthal bone samples showed that Neanderthals were a separate evolutionary lineage from modern humans and that, within the limits of the mtDNA analysis, they constituted a dead-end branch that, *it was assumed at the time*, did not contribute to contemporary human populations. Neanderthals and modern humans split from a common ancestor around 600,000 years ago. The ancestors of Neanderthals then spread to Europe, the Middle East, and Central Asia, where they are thought to have coexisted with humans for around 5,000 years after the Out of Africa migration.

Several years later, after the development of new and powerful genomic technologies, such as next generation sequencing (NGS, discussed in Chapter 5), it became possible to sequence the entire genome of several different Neanderthal bones. These genome sequences, although overall very similar to those of modern humans, contained specific regions that differed enough to distinguish these two closely related but genetically distinct subspecies. As Paabo and his colleagues continued to generate genome sequences from Neanderthal remains and compare these to modern contemporary human genomes, they noted that many humans, in particular those from Europe and Asia, appeared to contain short stretches of DNA with the characteristic Neanderthal sequences, comprising 2%–4% of the total genome. This remarkable observation suggested that there had, in fact, been some interbreeding between Neanderthals and the ancestors of modern humans. Of course, the media and general public were, quite understandably, intrigued with the idea that our ancestors had sex with Neanderthals and that, many thousands of years later, our genomes still retained traces of this genetic legacy. The evolutionary contribution from these archaic subspecies found in modern human genomes has been termed "introgression." The multiple introgression events whose traces we can detect in modern human genomes are estimated to have occurred 47,000–65,000 years ago.

The initial inference about interbreeding is based on a simple logical argument involving the geographic distribution of DNA sequence variation and on the now well-accepted Out of Africa migration model. The 2%–4% of genome sequences attributed to Neanderthals found in contemporary human genomes was observed only in individuals in Europe and Asia and, crucially for this argument, *not in Africa*. The simple Out of Africa model based on a bottleneck migration into Europe and Asia predicts that the rest of the world (ROW) has less genetic variation than Africa. A geographic pattern in which Neanderthal DNA sequences are present in some small part of modern human genomes outside Africa (Eurasia) suggests that they arose *after* the Out of Africa migration and could have come from interbreeding between the archaic Neanderthals in Eurasia and the ancestors of modern humans when and where these two subspecies coexisted. An alternative explanation is that these presumed Neanderthal sequence variants actually reflect ancestral polymorphism, transmitted to both Neanderthals and modern humans, and were initially present in modern human populations in Africa but, for some reason, disappeared from Africa. This explanation seems more complicated, requiring additional assumptions, and is certainly much less interesting than the inference of interbreeding between Neanderthals and our modern human ancestors.

Sequencing genomes from several different Neanderthal fossil remains confirmed that this well-known population found in Europe and Asia was a distinct human subspecies. Shortly thereafter, DNA analysis identified yet another archaic subspecies. In the 1970s, a group of Russian archeologists had found a variety of fossilized bones in a cave in Denisova, Siberia. Following the dramatic reports of DNA sequences from Neanderthal bones, they sent samples from the Denisova cave to Paabo and his colleagues at the Max Planck Institute for Evolutionary Anthropology. The Max Planck researchers were able to extract DNA from some of these bones. In 2010, they sequenced these DNA samples and identified most of them as Neanderthal; however, the sequence from a tiny fingerbone suggested another previously unknown archaic subspecies. The DNA sequences from this fingerbone turned out to be distinct from those of Neanderthals as well as modern humans and, so, a new archaic subspecies, the Denisovans, was *defined solely by DNA sequence* in the absence of a fossil record (Reich et al., 2010). For many years, this fingerbone remained the only evidence for this newly identified human subspecies. In 2018, far from the Denisova cave, on the high-altitude Tibetan Plateau, researchers excavated the Baishiya Karst Cave, Xiahe, Gansu, China, and found several fossilized remains (Chen et al., 2019). One was a human-like mandible, the lower jawbone, estimated to be at least 160,000 years old. Attempts to extract DNA failed but an ancient protein could be analyzed.[4] The limited amino acid sequence of this protein and the inferred DNA sequence turned out to be more similar to Denisovans than to Neanderthals or modern humans.

We know much less about Denisovans than we do about Neanderthals. In the absence of an extensive fossil record, we don't know what they looked like but we can estimate that they overlapped in Asia with Neanderthals for thousands of years. And, based on a remarkable 90,000-year-old bone fragment found in a Siberian cave, we now know they interbred. When Viviane Slon, a graduate student in the Max Planck Institute analyzed the mtDNA from this sample, Denisova 11, the sequence turned out to be

---

[4] The sequence of bases in DNA can predict, via the genetic code that specifies which three bases encode a given amino acid, the sequence of amino acids in a given protein. Using the same code, the DNA sequence can be inferred from the amino acid sequence.

clearly Neanderthal. But when the nuclear DNA fragments were analyzed and the genomic sequence was determined, she and her colleagues were surprised to find a mixture of Neanderthal and Denisovan sequences. For each pair of chromosomes, one came from a Neanderthal and the other from a Denisovan; the presence of two X-chromosomes indicated a female. They concluded that this Denisova 11 bone fragment came from a female hybrid whose mother was Neanderthal, based on the mtDNA, and whose father was Denisovan (Slon et al., 2018).

Denisovans interbred not only with Neanderthals but also, like Neanderthals, with the ancestors of modern humans. DNA sequences characteristic of Denisovans have been found throughout East Asia, reaching 4% of the genome sequences in Aboriginal Australians and Papuans. A Philippine ethnic group known as the Avta Magebukon has the highest known level (5%) of Denisovan ancestry in the world. According to Chris Stringer, a professor and research leader in Human Origins at the Natural History Museum in London, these findings suggests that different Denisovan populations mixed and intermingled with *Homo sapiens* in multiple locations and at various points in time and may have interacted more widely with our ancestors than did Neanderthals. "Relatively small groups of early modern humans interbred with Neanderthals in western Eurasia and then spread across Eurasia and beyond, passing on that level of acquired Neanderthal DNA to descendant populations," Professor Stringer pointed out. "In the case of the Denisovans, it looks like they were genetically much more diverse, and they intermixed separately in different locations with differentiating early modern populations, hence the more varied patterns we see today." Professor David Reich of Harvard University, one of the leaders in the ancient DNA field, estimated that the human–Denisovan admixture took place after the one between modern human and Neanderthals. Reich, as well as other ancient DNA researchers, has pointed out that some of the genetic variants modern humans have inherited from the archaic subspecies with whom they interbred have helped them adapt to new and challenging environments. The best-known example of such "adaptive introgression" is a variant of the EPAS1 gene that regulates the response to hypoxia (low oxygen), present in Tibetan populations living on the high altitude Tibetan Plateau and derived from our Denisovan ancestors. Statistical analyses of DNA sequences can, as we've seen using the "molecular clock," provide estimates of time; they can also reveal evidence of positive Darwinian selection for adaptation. These analyses indicate strong positive selection for this

EPAS1 variant we (or, at least some of modern humans) have inherited from the Denisovans.

A recent excavation in Chagyrskaya Cave in the foothills of the Altai mountains in southern Siberia, a mere 100 meters from the now famous Denisova cave, has identified the largest collection of Neanderthal bones found in a single cave. DNA sequencing of this large group of bone fragments and teeth, all dated to between 50,000 and 60,000 years ago, almost doubled the number of Neanderthal genomes available for study. Analysis of the sequences revealed a father and his daughter and two other more distant relatives as well as two more from a nearby site, providing insights into the social structure of this community (Skov et al., 2022). For one pair of remains, half of the genome sequence was shared, suggesting that they were father and daughter or siblings. Their mtDNA sequences did not match, inconsistent with a sibling relationship. Other familial relationships were revealed by analysis of the mtDNA. The father's mtDNA sequence had a "heteroplasmic site," a position in the mitochondrial genome where two different bases were present. (Mitochondrial heteroplasmy is discussed in more detail in Chapters 4 and 8.) Two other adult males shared this mitochondrial sequence, including the heteroplastic site, indicating that they were from the same maternal lineage. Since heteroplasmies usually disappear after a few generations, with one of the two sequences becoming dominant, Skov inferred from the patterns in the three adult males that they all lived around the same time.

The investigators also noted that the genetic diversity of the mitochondrial sequences was greater than that of the Y chromosomes, suggesting that the females in this Neanderthal community had moved from elsewhere to live in the male's community, a "patrilocal" pattern.[5] Also, the genome sequences of the Chagyrskaya community proved to be more similar to Neanderthals present in Europe at the same time than to the Neanderthals living in the nearby Denisova cave thousands of years earlier. So, like modern humans, Neanderthals migrated long distances, often replacing the local communities. I continue to find it astonishing how much of human history can be revealed from simply analyzing DNA sequences.[6]

---

[5] In a matrilocal pattern of migration, males move to live with the female's community. Analyzing the genetic diversity of mtDNA and of Y chromosomes has identified both of these cultural patterns in different modern human tribes, although patrilocality is more common.

[6] Analyzing sequences from nonhuman sources, such as the common human parasite the head louse (*Pediculus humanus capitis*) and the closely related body louse (*Pediculus humanus humanus*), can also illuminate critical aspects of human history. Based on a phylogenetic analysis of these sequences, Mark Stoneking noted that the body louse, which lives in clothing, appears to have evolved from the head louse around 75,000 years ago. He proposed that the custom of wearing

As we've seen, analyzing the genome sequences of archaic species like the Neanderthals and Denisovans can reveal critical details of our ancient past. By comparing them to the genomes of modern humans, we may also gain some insights into what makes us uniquely "human." Neuroscientists have demonstrated that the frontal lobe of humans has many more neurons than that of chimpanzees and that, based on studies of fossilized hominin skulls, the brains of our ancestors increased in size, reaching the size of modern humans around 600,000 years ago. The *size* of the Neanderthal brain is essentially the same as that of modern humans, but there are differences in the *shape*. Now, we have the opportunity to compare the potentially relevant genes of our closest extinct hominin relative with our own. Championed by Svante Paabo, this program of comparing archaic to modern human genomes has revealed that, of the 20,000 or so protein-encoding human genes, most are identical to those in Neanderthals and Denisovans but human-specific mutations in 96 different genes have changed the structure of a protein.[7] The change in one or more of these genes may account, in part, for the presumed differences in brain function between modern humans and these archaic species.

In 2017, Anneline Pison, a researcher in the laboratory of Wieland Huttner and neuroscientist at the Max Planck Institute of Molecular Cell Biology and Genetics in Dresden, Germany, noticed that one of these mutations changed the gene TKL1, known to be active in the developing human frontal lobe. In a series of remarkable and elegant experiments, Dr. Pinson and her colleagues demonstrated that the human variant of TKL1 gave rise to more neurons than the Neanderthal variant (Pinson et al., 2022). After injecting the human variant into the developing brains of mice and ferrets, they observed an increase in the growth of neurons. Subsequently they edited the TKL1 gene in a human embryonic stem cell to create the TKL1 variant found in Neanderthals and chimpanzees and then induced the stem cells to develop into brain tissue, called a brain organoid. This Neanderthal-like brain organoid generated a miniature cortex with fewer neurons than the organoid with the human TKL1. Functional analysis, like this series of experiments,

clothes originated at that time, creating an new environmental niche to which the new louse subspecies adapted (Kittler, Kayser, and Stoneking, 2003), noting that this time frame corresponds to the estimated time when modern humans migrated out of Africa. Mark told me he was inspired to conduct this study when his son came home from school with head lice.

[7] An excellent recent review of human-specific mutations compares the genome sequences from modern and archaic hominins, great apes, and other primates (Pollen et al., 2023).

for some of the other genes on this list of 96 human-specific mutations may prove to shed more light on how the complex cognitive functions of modern humans have evolved.

Thus far, most of the studies of archaic humans and of ancient DNA have been performed on bones but, as it turns out, there are very few fossilized archaic remains available for analysis. Even with the recent discoveries in the Chagyrskaya Cave, there have been only around 20 different Neanderthal whole genomes identified, and only one Denisovan sample has yielded DNA sequences. The desire to overcome the limited availability of fossilized remains led to an innovative and, on its face, unlikely strategy. Based on the assumption that there might be ancient DNA present in the caves in Europe and Asia where the fossils or tools of archaic humans had been identified, a team led by Viviane Slon and Matthias Meyer of the Max Planck Institute attempted using PCR to amplify DNA found in the dirt at these sites. I had the opportunity to hear this work presented at a conference in Croatia in 2017.[8] Like most of the audience, I was amazed and impressed that this strategy had actually led to the identification of some DNA sequences that appeared to be derived from a variety of mammals, including archaic humans. Focusing on DNA sequences with chemical damage characteristic of ancient DNA, the team recovered nine different Neanderthal mtDNA (partial) genomes from four different cave sediments. They also were able to find Denisovan mtDNA from the Siberian cave where the famous fingerbone that first identified this archaic group had been found. A rough date for the DNA could be estimated based on where it had been found in the sediment layer pattern. While the recovered DNA fragments were short and only mtDNA was identified, that anything at all could be amplified and sequenced from the dirt in these caves was remarkable.

How is this feat of amplifying the DNA present in a dirt sample achieved? As we've seen in Chapter 5, a "shotgun" library can be created by attaching short synthetic DNA fragments (oligonucleotides) of a defined sequence using the appropriate enzymes to both ends of whatever natural DNA fragments are present in the sample. Then, PCR primers whose sequence is complementary to these attached synthetic DNA sequences can be used to carry out PCR amplification, creating a "library" of the DNA in the sample,

---

[8] The International Society for Applied Biological Sciences (ISABS) conferences are held in Croatia every two years and focus on medical genetics, forensic genetics, and anthropological genetics. My research has also focused on these apparently disparate areas; the latter two are the subject of this book.

ready for sequencing on an NGS instrument. In principle, this strategy confers "immortality," the ability to be replicated by PCR amplification, on all of the DNA fragments present in the sample.[9] The amplified DNA fragments in the library can then be sequenced and analyzed to try to figure out their source. Similar strategies have been used to discover many new species of marine micro-organisms present in a drop of ocean water. In the nomenclature tradition, illustrated by qPCR for quantitative PCR, this strategy of identifying a new species in the environment is known as ePCR.

This nonselective application of PCR is fundamentally different from our initial *targeted* amplification of the beta-globin gene from human genomic DNA back in the mid-1980s. We were using PCR to amplify only a tiny *specifically targeted* segment of the human genome; the portion of the genome we targeted, the beta-globin 110 bp fragment, was around 1/10 million, the classic "needle in the haystack." Nonetheless, the amplification of whatever is present in the sample (the shotgun library) and the targeted PCR amplification of a specific DNA fragment sequence share the same basic biochemical procedure with repeated cycles of PCR with primers and DNA polymerase. In the case of the archaic DNA recovered from dirt, however, the creation of a shotgun library was modified by the Max Planck group to accommodate these short DNA fragments, much of which consisted of single DNA strands rather than the well-known double helix. In addition, they used synthetic DNA fragments ("capture probes") complementary to mtDNA to "capture" or "fish out" the tiny amounts of ancient mtDNA present in the DNA amplified from the cave sediment.

Sequencing ancient DNA from dirt was clearly a technical tour de force. Two years later, at the 2019 ISABS conference in Croatia, I heard a remarkable *conceptual* breakthrough. Josh Akey (Princeton) described a strategy he termed "fossil-free sequencing of archaic genomes in modern DNA." Akey noted that it's very difficult to sequence the limited and degraded DNA present in fossilized remains . . . and there are so few of them.[10] Given that many modern humans have a few percent of Neanderthal- or Denisovan-derived DNA in their genomes, why not analyze archaic DNA by sequencing the genomes of contemporary humans? The genetic diversity of archaic DNA

---

[9] Of course, this *potential* "immortality" is realized only when some entity (human or robot) adds PCR primers, DNA polymerase, and the dNTP building blocks to amplify the DNA fragments.

[10] This description reminded me of the ancient comedic complaint: *the food is terrible . . . and the portions are so small.*

may be more effectively revealed by sequencing these genomes than the relatively limited number of fossilized remains.

While the sequencing of ancient Neanderthal and Denisovan fossil remains has provided the most dramatic narratives of ancient DNA research, the analysis of ancient human bones has also provided novel insights into our history. As we've seen, modern humans migrated out of Africa around 75,000 years ago and expanded into Europe and Asia and, eventually, to all the other continents. A very recent paper (Posth et al., 2023) analyzed the genome sequences of 357 ancient Europeans, the oldest dating back 45,000 years, and proposed that several different waves of hunter-gatherers had migrated into Europe, comprising 8 distinct populations. The genetic diversity of these groups was, surprisingly, greater than among modern Europeans and Asians. These groups coexisted in Europe for thousands of years, but one of these groups, the Vestonice, the oldest and genetically the most distant, did not survive the Ice Age. So, some of the first modern humans to settle in Europe shared the fate of the Neanderthals and were eventually replaced by other hunter-gatherer groups.

Another recent study of some human remains from the Middle Ages sheds light on the social history of medieval England. In 2004, construction workers in Norwich, UK, uncovered the human skeletal remains of 6 adults and 11 children at the bottom of a medieval well. Many years later, these individuals were shown to be Ashkenazi Jews, and 4 of them were closely related, based on NGS genome sequencing analysis and historical documents (Brace et al., 2022). The skeletons were all positioned head first, suggesting to the researchers that this group of 17 people could have been the victims of an anti-Semitic massacre and deposited in the well.[11] Radiocarbon dating indicated that the remains were from the late 12th or early 13th century, a period of well-documented anti-Semitic violence in England, such as the Norwich massacre of 1190 CE.

The genome sequences revealed that these individuals carried four different genetic variants (alleles) associated with genetic diseases, such as Tay-Sachs, common among modern Ashkenazi Jewish populations. Computer simulations showed that the frequency of these disease mutations in the medieval Ashkenazi population would have been similar to those in the modern population. These disease mutations are thought to have risen to their

---

[11] The discovery of these remains at the bottom of a well brings to mind the faux folk song "Throw the Jew Down the Well," written and sung by Sacha Baron Cohen, as his creation Borat, the faux Kazakh journalist. Little did Cohen know that his dark satire reflected such a grisly medieval reality.

current frequency as a result of a population bottleneck in the past, when certain alleles could increase in frequency by chance due to the reduction in population size. These results implicate a bottleneck *prior* to the 12th century, earlier than previous hypotheses, which had dated the event to around 500 to 700 years ago.

Another study in a medieval Jewish cemetery in Erfurt, Germany, extracted and sequenced DNA from the teeth[12] of 33 men, women, and children, comparing the sequences with those of modern Jews (Waldman et al., 2022). The genomes of these medieval Jews carried the same disease-causing mutations at similar frequencies as modern Ashkenazi Jews. Combined with the data from the 17 individuals at the bottom of the Norwich well, these results support the idea of a very early population bottleneck. In addition, these genome sequences revealed some unexpected differences with those of modern Ashkenazi Jews. The individuals buried in the Erfurt cemetery had two distinct genetic lineages and were more diverse than modern Ashkenazi populations. How the modern populations became more homogeneous remains something of a mystery.

Sequencing the genomes of ancient human remains has also provided detailed insights on the peopling of the Americas. Prior to the ability to sequence the genomes of fossilized remains, the work of many labs, including our own, as I discuss in Chapter 11, indicated that the ancestors of Native Americans migrated from Siberia, crossing the Beringian land bridge and subsequently populating North and South America. These reconstructed histories were based on analysis of population data, namely, the frequencies of genetic variants in different population groups. The reduced genetic diversity of the Native American groups indicated that this "Out of Siberia" migration involved a major population bottleneck or "founder effect." How these first settlers spread through the Americas and the timing of these migrations, however, remain unclear, questions that analysis of DNA sequences can help illuminate. The recent analysis of the remains of an infant who lived 11,500 years ago in Upward Sun River in Alaska (dubbed USR1) by Eske Willerslev (University of Copenhagen) and his colleagues (Moreno-Mayar et al., 2018) sheds valuable and much-needed light on how this founding population from Beringia spread southward. Comparing the genome

---

[12] DNA was extracted from teeth rather than bone fragments because a rabbinical court judge in Israel had ruled that it would be permissible to extract DNA from teeth, which, unlike the rest of the skeleton, do not require reburial under Jewish law.

sequence of USR1 to a set of ancient genomes as well as to a panel of 167 worldwide populations, genotyped at around 200,000 SNPs, demonstrated that USR1 was most closely related to Native Americans, followed by Siberians and East Asians. The USR1 sequence appears to be ancestral to all previously sequenced contemporary and ancient Native American samples and represents a distinct "Ancient Beringian population." Since comparing sequences allows for calculation of divergence times, Willerslev and colleagues were able to estimate that the founding population of Native Americans diverged from ancestral Asians around 36,000 years and subsequently, Ancient Beringians diverged from the common ancestor of other Native Americans groups around 20,900 years ago. These time estimates suggest that the population groups from Asia that first migrated to Beringia stayed there for a significant time before moving southward and settling the Americas. The split between Northern and Southern Native American populations was estimated to occur around 15, 000 years ago.

For some Native American groups, analyzing ancient DNA has clarified disputed claims of ancestry and resolved some contemporary conflicts. I live in the San Francisco Bay area, where the Ohlone people originally lived on 4.3 million acres. The Muwekma Ohlone Tribe, a descendant community, has been trying since 1989 to regain federally recognized status, which they lost in the 1920s when the well-known UC Berkeley anthropologist Alfred Kroeber claimed that the tribe was "culturally extinct." Federal recognition provides legal sovereignty and access to federal programs for tribal support. Since the 1920s, when they numbered fewer than 100 members, the Ohlone have survived and now have 500 tribal members. Their ongoing fight for federal recognition uses genealogical records and legal documentation to demonstrate a long history of their presence in the Bay Area. Now ancient DNA joins the struggle.

In 2016, two indigenous village sites near Fremont, in the San Francisco Bay area, were uncovered during a construction project. Radiocarbon dating by a team of archeologists indicated that one of the sites was occupied between 490 CE and 1775 CE and the other between 1345 and 1850 CE. Their location suggested that the modern Muwekma-Ohlone were the most likely descendants of the individuals whose remains were found at these two sites. Working with members of the modern Muwekma-Ohlone, who contributed genealogical and oral histories, the geneticist Ripan Malhi (University of Illinois, Urbana-Champaign) was able to extract DNA and recover genome sequences from the remains of 12 different individuals as well as from 8 living tribe members. Malhi and his colleagues compared these 12 ancient genome sequences to genomes of indigenous individuals from throughout the Americas and found that they were most similar to ancient individuals from Southern California. Specific distinctive genome sequences from the individuals in these two sites were also shared with modern tribal members, "providing evidence of genetic continuity between past and present individuals in the region, in contrast to some popular reconstructions based on archeological and linguistic information" (Severson et al., 2022). The older history based on linguistic analysis proposed that the Ohlone first arrived in the Bay area from the north around 1500 CE. The similarity between the DNA sequences from the Fremont sites and those from the ancient DNA samples from Southern California indicates that the Ohlone came to the Bay area around 1,000 years earlier and from the south rather than the north, consistent with the Ohlone's own version of their history.

Efforts to restore the Ohlone's tribal status are underway at the federal level; the restoration of Ohlone culture is already happening. In a supreme irony, the newly relocated Café Ohlone, a gathering place that celebrates indigenous cuisine and culture, now sits in the courtyard of the UC Berkeley's Hearst Museum of Anthropology, an institution that houses many of the Ohlone remains, cultural artifacts, and relics, and whose director from 1908 to 1946 was none other than Alfred Kröber. In the case of the Ohlone, the genetic similarity between the 2,000-year-old remains and the modern tribe members indicates that the bones in the two village sites are from the ancestors of the modern Muwekma-Ohlone.

This relationship between the genomes of those below and above the ground is not always the case. Comparing DNA data from contemporary populations with ancient DNA from the same region tells us that human history is often one of replacement as well as of migration. As David Reich (2018) and other ancient DNA researchers have pointed out, for any particular region, the ancestors of the current inhabitants may not be the ancient bones buried beneath them, but more recent migrants from elsewhere.

This Just In

I began writing this book over 20 years ago; it has been simmering on a back burner while writing grants and scientific journal articles took precedence. In the course of such a long gestation, new technologies and applications arise, with new discoveries upending previously held assumptions. In the beginning of this chapter, I discuss the 1993 *Nature* paper by Tomas Lindahl, which explained why recovering DNA from samples that were a million years old was highly implausible due to the degradation of DNA over time. DNA becomes fragmented due to the enzymatic activity of micro-organisms, mechanical shearing, and chemical reactions, such as oxidation and hydrolysis. Now, in the winter of 2022, a *Nature* paper by the group of Eske Willerslev (Kjær et al., 2022) reports the discovery of environmental DNA from the permafrost of Greenland that is at least 2 million years old. For several years, Willerslev's lab at the University of Copenhagen has been refining the technology of making shotgun libraries and sequencing the very short DNA fragments found in the Kap Kobenhavn Formation, a 100-meter-thick sediment deposit at the mouth of a fjord in North Greenland. The short DNA sequence reads generated by NGS analysis of the nuclear and mtDNA in the libraries were aligned with both plant and animal sequence databases to identify the source of the environmental DNA. The age of the DNA fragments was estimated by geologic dating of the sediment layers and radiocarbon dating of some specimens, as well as the sequence-based time estimates using the molecular clock. The authors proposed that the unexpected survival of these DNA fragments may have been due to their binding to mineral surfaces and speculated that "adsorption at mineral surfaces modifies the DNA conformation, probably impeding molecular recognition by enzymes, which effectively hinders enzymatic degradation."

The many different plant, animal, and marine species, revealed by analysis of the environmental DNA, suggests a diverse boreal forest ecosystem. The mtDNA indicates a variety of mammals, including reindeer, caribou, rodents, geese, and mastodons, . . . all ancestral to their present-day relatives, based on the DNA sequences. Detecting marine species, like the horseshoe crab, indicates a significantly warmer temperature than present-day Greenland. Although this landmark paper, unlike the studies discussed earlier, does not address the issue of human evolution, it paints the picture of a 2-million-year-old unexpectedly vibrant and diverse ecosystem in what is now a polar desert in northern Greenland. It also illustrates how far the field has come and how rapidly the DNA technology has evolved since that first 1984 report on the quagga mtDNA.

11

Populations, Genes, and History

We now know, based on the genetic data discussed in the previous chapters, notably the analysis of mitochondrial (mt) DNA sequences by the Wilson group, that all modern humans emerged from Africa around 75,000 years ago,[1] moving into Europe and Asia and, then over time, populating the rest of the planet. This broad global pattern of migration, based on the DNA analysis of modern populations and ancient remains as well as on archeological findings, is illustrated in Figure 11.1. How did this this remarkable diaspora occur? How are all the different populations related to one another? And what kind of data can help us reconstruct the details of these historical migrations, including the relatively recent peopling of the Pacific islands (around 2,500 years ago) and the Americas (around 20,000 years ago)?

Although some of this pattern was inferred from the analysis of DNA sequences, these earlier pictures of human migrations were based primarily on population data, namely, the frequency of genetic variants (alleles) in different populations. Over 25 years ago, our lab observed that the frequencies of many HLA alleles found in Native American groups are most similar to those of populations in Siberia, suggesting a migration from Northeast Asia to the Americas. Another example is a study of the origin of the Sami, a native population living in the north of Sweden and Norway with a distinct culture and language. Ulf Gyllensten, a professor at the University of Uppsala and a friend and frequent collaborator, collected samples while helping to provide medical services to these communities. Our lab, notably Steve Mack, carried out the HLA genotyping and statistical analysis. What we found was that many of the common HLA alleles in the Sami were present in European populations but that several common HLA alleles were very rare or absent in Europe but common in Asian populations. Our analyses indicated that 87% of the Sami gene pool is of European origin and that the Asian contribution is around 13%, suggesting that some of the ancestors of the current Sami

---

[1] Estimates for the Out of Africa migration vary and, of course, modern Africans represent populations that did not leave Africa or participated in back-migrations.

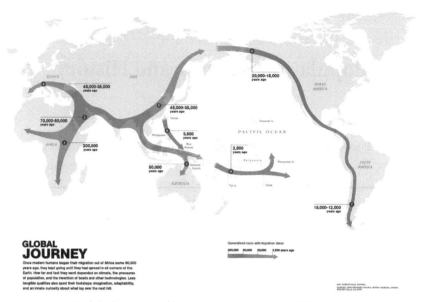

Figure 11.1 A global pattern of human migrations (modified from *National Geographic*).

population migrated to northern Scandinavia from Asia. The broad global migratory pattern illustrated in Figure 11.1 is, of course, based on more recent studies from many different labs using many genetic markers with many different populations.

In Part I, we saw that population data describing the allele frequencies for a polymorphic marker are required to interpret the significance of a match between the evidence specimen and the suspect. Estimating the probability of a purely coincidental match (the random match probability, or RMP) based on the population frequency of a specific genetic profile is one way of interpreting a forensic match. (Likelihood ratios, the other commonly used forensic metric, are also related to population frequency data.) As we see in Figure 11.1, the very same kind of population data can be used to examine the relationships among different human populations, to make inferences about human migrations, and to test hypotheses about the recent history of human populations.

In the early 1970s, the Italian geneticist Luigi Luca Cavalli-Sforza and his colleague A. W. F. Edwards had the insight that analysis of the allele frequency differences between different human populations could be used to generate phylogenetic trees or networks that reflected the genetic relatedness of the populations in question.[2] Cavalli-Sforza, whom I got to know during his time at Stanford, applied this fundamental insight to consider specific historical questions. One of the first issues he addressed was how agriculture spread from the Middle East to Europe and the rest of the world. The evidence for the diffusion of farming and domesticated plants was based on the pattern of cereal remains (carbonized grains and grain impressions) observed at Neolithic sites. Based on his analysis of the frequency of genetic markers in various Middle Eastern and European populations, he proposed a migration out of the Middle East region where farming is thought to have originated (the "fertile crescent") into Europe. He concluded the most probable explanation for the historical expansion of agriculture was that "it was not the *idea* of farming that spreads, but the farmers themselves." This initial analysis was based not on the DNA markers that were emerging during the 1980s but on blood groups and other blood based polymorphic systems, including the HLA markers (in their pre-DNA versions). The subsequent availability of hundreds of thousands of DNA markers and analyses of the allele frequencies from different populations have made these trees and networks much more detailed and, from a statistical perspective, much more reliable.

The basic idea underlying the construction of these population trees is to use the frequencies of the genetic markers in each population to create a network connecting them all. For every possible pair of populations in the study, the difference between allele frequency distributions was calculated. This difference is termed "genetic distance." A network or population "tree" connecting all the individual populations could then be constructed from these pairwise genetic distances by a statistical algorithm (Figure 11.2). These population trees look similar to the phylogenetic trees based on DNA sequences discussed in the previous chapters but differ in a few fundamental ways. In population trees, the individual datapoints or *taxa* in the tree are population frequency distributions based on hundreds or thousands of genetic markers and of *populations* of individual DNA samples rather than on *individual DNA sequences*. For the construction of phylogenetic trees of DNA sequences, no subjective judgments are required for defining the *taxa* (the individual DNA sequences). Constructing trees from population frequency data, however, raises the question of how these population groups are defined. Should, for example, indigenous populations from Australia be separated into the three tribal groups shown in Figure11. 2 or combined as one group of Australian

---

[2] Reviewed in Cavalli-Sforza, Menozzi, and Piazza (1994).

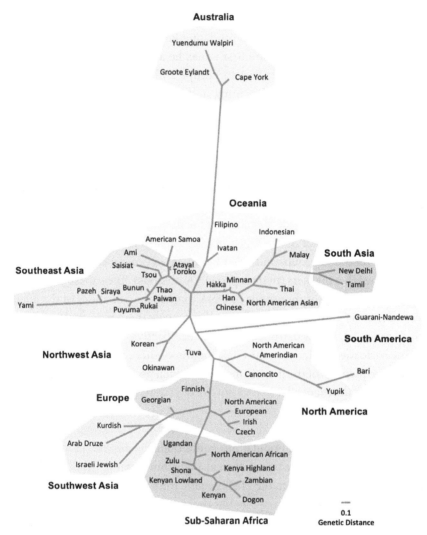

Figure 11.2 A population network of 52 global populations based on the HLA-A-B haplotype frequencies (Mack and Erlich, 2006).

Aborigines? Should the population groups from Kenya be separated into Highlands and Lowlands? Defining these population categories for analysis is obviously more subjective than identifying an individual DNA sequence.

What can be inferred from these two types of trees is also very different. The relationship between sequences in a phylogenetic tree can be understood to reflect discrete base changes (mutations) from an *ancestral* to a *derived*

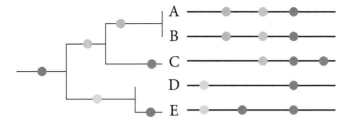

Figure 11.3 Phylogenetic trees of DNA sequences can reveal the history of accumulated mutations

In this simplified picture, the sequence with the blue dot represents the ancestral sequence, with the other colors representing the accumulation of mutations, giving rise to five derived sequences (A, B, C, D, E).

sequence and interpreted as a historical progression. A specific sequence of mutational events can be reconstructed to link the various DNA sequences (see Figure 11.3).

The interpretation of the population tree or network is less clear; populations that cluster are more closely related (shorter genetic distances) but the reasons for the patterns of clustering are not as well defined or well understood. Nonetheless, the patterns of population relatedness revealed by these networks have been widely used to infer historical migrations.[3]

The population network shown in Figure 11.2, illustrates the genetic relationships among 52 different worldwide human populations, constructed by my colleague and long-time collaborator Steve Mack, based on population frequency data for two highly polymorphic HLA genes. As noted in previous chapters, the HLA genes are the most polymorphic in the human genome and are therefore highly informative in population genetic analyses and in reconstructing human migrations. These are also the genes that govern the immune response and are involved in transplant donor and recipient matching; one of these genes, the HLA-DQA1 marker, served as the first forensic DNA test. The population data underlying this network was generated as part of an international collaboration (The 12th International Histocompatibility[4] Workshop) examining how HLA diversity was

[3] The population data displayed here as unrooted networks can also be displayed as trees or dendrograms. My preference is for the network rather than the tree display because the root and branching structure of the tree implies a more specific history . . . , that is, population A "split" and gave rise to population B and C . . . than does the more "open" structure of the network.
[4] The homologues of the HLA genes, known as the human major histocompatibility complex genes, were first defined in mice by skin graft experiments, hence the term "histocompatibility." For

distributed around the world. The power of these analyses is always limited by the number of populations examined and the number of genetic markers used. Subsequent analyses with more populations and many more genetic markers provided a more detailed reconstruction of the patterns of human migrations, but this network based on only two HLA genes still yielded some valuable insights. In light of the recent analyses of ancient genomes using new DNA sequencing technology and sophisticated statistical methods, these early studies seem pretty rudimentary. In a sense, our studies with HLA markers represent the "ancient past" of the ongoing search to understand human history using DNA analysis.

Phylogenetic trees of human DNA sequences can use an "outgroup" (e.g., chimpanzee sequences) to define a "root" to establish *directionality*, as shown in the mitochondrial DNA tree in Chapter 9. There is no obvious "root" to this HLA population network, so we have to make some assumptions about the historical movement of the various populations depicted here. The directional interpretation of this "unrooted" network is based on the overwhelming amount of paleontological and genetic evidence that all modern humans emerged out of African around 75,000 years ago.[5] In general, the genetic relationships reflect geography, with populations within a geographic region (e.g., sub-Saharan Africa, Europe, etc.) clustered together. The history of human migrations inferred from these limited population data is a passage out of the Old World (Africa, the Near East, Europe) into Asia and then to the more recently populated regions of Australia, Polynesia, and the Americas. The length of the lines separating populations does not indicate *geographic* distance between groups but the difference in allele frequencies (*genetic* distance).[6] The observation that the populations most closely related to African populations are Europeans and the Middle Eastern groups might seem surprising to people focused on traits like skin color, but certainly not to those familiar with maps of the region. For the more recently colonized regions, such as the Americas and Polynesia, some of the population clusters suggested specific migratory patterns. The Tuva, a Siberian population, was closest to the cluster of Native American populations consistent

---

this study, Steve sent out the PCR-based HLA genotyping kits we had developed to around 50 different labs around the world, curated the data, and analyzed the results (Mack and Erlich, 2006).

[5] As previously noted, modern Africans are descended from populations that remained in Africa. Some very recent genetic studies suggest that there may also have been an earlier migration out of Africa as well as some back migration from Europe into Africa.
[6] The length of the branch can also reflect isolation time and genetic drift.

with the long-held idea that the Americas were colonized by Northeastern Asian groups migrating over the Beringian land bridge to what is now Alaska and then spreading throughout the Americas. In a less obvious relationship, a Polynesian population, American Samoa, clustered with a geographically distant group of indigenous Taiwanese populations. Steve Mack and I pursued the question of how Polynesia was colonized with a subsequent study with many additional Pacific island populations (Figure 11.4).

In general, then, the structure of the population genetic networks correlates with geography. Populations that are geographically close are more likely to have a common ancestral population as well as to have more interpopulation gene flow (aka mating) than do populations that are geographically distant.

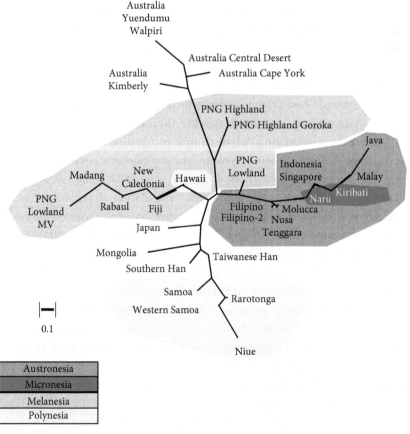

Figure 11.4 Population network based on HLA-DRB1 allele frequencies (from Mack et al., 2001).

So this correlation is *expected*—not a mysterious observation but a default assumption. When the patterns of these genetic networks *deviate* from geography and proximity, however, the data can be used to test different hypotheses about recent human history.

The Peopling of Polynesia

The spread of human populations across the vastness of the Pacific Ocean has been one of the most dramatic and awe-inspiring migrations in human history. The details of how the far-flung islands of Polynesia were settled was, until very recently, a mystery. Anthropologists had been debating two competing hypotheses about the colonization of Polynesia, thought to have occurred around 2,000 years ago. One hypothesis, the "Entangled Bank," proposes that groups from continental Asia migrated gradually over thousands of years to the islands of Melanesia, Micronesia, and Austranesia, with extensive interbreeding with local populations, eventually arriving over vast distances to settle the widely dispersed islands of Polynesia. The other, the oddly named "Fast Train to Polynesia," postulates that groups from continental Asia migrated relatively quickly through Melanesia, Micronesia, and Austranesia, colonizing Polynesia with minimal interbreeding with the local groups they encountered en route. The phylogenetic network generated from the population data (here from just one HLA gene, HLA-DRB1) (Figure11.4) shows that the Polynesian population datapoints (Samoa, Western Samoa, Raratonga, Niue) are located closer to those of the continental Asian populations (Chinese Han, Taiwanese Han, Mongolia) than to the geographically neighboring populations of Melanesia, Micronesia, and Austranesia, which is consistent with the predictions of the "Fast Train" hypothesis. The consensus now, more than 25 years after our early studies with HLA markers, is that the spread of Polynesian culture started with a migration from Taiwan. As we noted previously, the American Samoa population clusters with a group of indigenous Taiwanese populations (Figure 11.2).

The pattern of this network also addresses another anthropological hypothesis about the history of human migrations in this region. In this network, the population datapoints in the Papua New Guinea (PNG) highlands are clustered more closely with those of Australian Aborigines than with the neighboring populations of the PNG lowlands. This pattern is consistent with the Sahul hypothesis that proposes the initial colonization occurred at a time during the last Ice Age, which ended around 12,000 years ago, when PNG and Australia were a single landmass, Sahul, and that the groups currently in the PNG lowlands resulted from later migrations *after* the waters had risen and separated what is now PNG and Australia.

Alternative Strategies for Analysis: Individual Genotypes versus Populations

All of these analyses of population-based networks were based on a definition of "population," of lumping together different individuals into discrete groups, raising the question of how many different groups should be defined and what kinds of distinctions (geography, language, religion, etc.) should be made. Should, for example, we lump together individuals from Europe, or as northern and southern European populations, by individual nations, or by subnational geographic groupings? If, for example, we had chosen to lump the Highland and Lowland populations from PNG into one group, we would have missed the narrative about the colonization of PNG and Australia inferred from the data in in Figure 11.4.

The groupings we ultimately choose influence not only our analysis of human history but also the way we consider the significance of the genetic profile matches between evidence and suspect discussed in Part I. The frequency of a specific genetic profile might be higher in some populations than in others and so the RMP could vary, depending on the population database used for the calculation. The US forensic population databases are defined very broadly and based more on the convenience of federal census categories than on genetics: Caucasian, Hispanic, African American, Asian, Native American, Pacific Islander. The Hispanic category is particularly problematic in terms of genetics as this group includes individuals with European, Native American, and African ancestry. How then, should the significance of a match be presented in the courtroom, given genetic profile frequency differences in these different population groups? One strategy is to provide the jury with RMPs calculated for different databases; for example, the RMP might be 1/50 million in one population and 1/2 million in another. Another is to focus on the population groups present in the geographic area where the crime was committed.

In general, the forensic genetic markers used for individual identification have been chosen, to the extent possible, to *minimize* the frequency differences between populations so that the RMPs calculated from different population databases are not so very different. However, a small set of genetic markers whose frequencies *are* very different in different populations can be used to infer the ancestry of the individual who left a particular forensic sample. Identifying the possible ancestry of an unknown perpetrator can serve as a valuable investigative lead, as discussed in Part I. These markers are known as AIMS (ancestry informative markers) and are used in their genetic analysis by commercial companies, like Ancestry.com and 23andMe, that provide estimates of ancestry based on analysis of cheek swabs or saliva samples and comparison to various ethnic reference population databases.

As we've seen, defining the relevant population groups for forensic analysis is complicated. Defining them for historical analyses is not straightforward either. One alternative to the construction of networks and trees based on population frequency data is to consider each individual genotype as a datapoint and use a statistical method known as principal component analysis (PCA) to cluster the individual datapoints. Here, the statistical algorithm creates the pattern of genetic relatedness *without any need to formally define population groups*. PCA is a method of simplifying the complexity in high-dimensional data, displaying the major trend(s) as a two-dimensional array.[7] It's an example of an unsupervised learning[8] method that can find patterns in the data without reference to prior knowledge or assumptions about whether the individual samples come from different groups.

As we see in Figure 11.5, based on using the PCA method, which focuses on the subset of genetic data that varies between populations, most of the individual samples fall into relatively discrete clusters, roughly corresponding to broad geographic population groups, with some individuals lying between these clusters. These individuals are genetically related to multiple population groups. In Figure 11.5, individual genotypes from European, African, and Native American populations form three discrete clusters; Hispanic individuals are arrayed either between Europeans and Native Americans or between Europeans and Africans.

How can the pattern generated by PCA, which appears to array individual genotypes based on biogeographic ancestry, be reconciled with the well-established observation that the vast majority of all genetic variation is between individuals *within* a population rather that *between* populations of

---

[7] Multidimensional PCA is also possible.
[8] This method, which sounds like a dangerous new age educational strategy for kindergarteners, is widely used for discerning patterns in complex data.

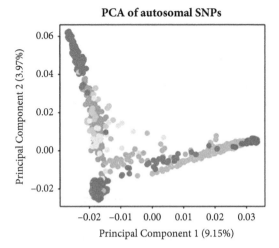

Figure 11.5 Principal component analysis (PCA) results of Hispanic/Latino individuals with Europeans, Africans, and Native Americans. Europeans are in red, Africans, in green, and Native Americans in blue (from Bryc et al., 2010).

different ancestry. This enormously important concept may be well established but is, nonetheless, not very clear to non–population geneticists. (I discuss and try to clarify this concept in Chapter 13 on race and genetics.) The explanation for this apparent discrepancy is simple. These principal component analyses focus on that very small amount of genetic variation capable of distinguishing and arraying these individual genotypes in a two-dimensional pattern. This pattern turns out to correspond to biogeographic ancestry. My point in discussing PCA is not this biogeographic correspondence but that this method can analyze genetic variation by considering individual genotypes without any prior assumptions about ancestral population groups.

DNA Sequences and Polynesia

DNA sequencing technology and statistical methods have both evolved significantly since our early studies with HLA markers, and the most recent picture of how Polynesia was settled is now much richer and much more detailed. Ancient DNA does not survive well in the moist tropical environment, but a new international study (Ioannidis et al., 2021) analyzed the genomes of 430 modern Polynesian samples from 21 different Pacific islands and was able to trace the sequence of migrations that settled the Polynesian island network.

The relationship between two different genomes can be estimated based on the sequence differences that have accumulated and on the length of the shared DNA sequences. (More distant relationships will have shorter shared DNA segments due to genetic recombination breaking up these segments over time.) Unlike most human migrations, the settling of Polynesia is basically a series of "founder events" in which a small population from one island migrates to other islands. Founder events are a particular kind of population bottleneck in which the number of different genetic variants in the island of origin is reduced, by chance, in the new settlement. To consider this pattern, the team developed a novel statistical method to trace the history of Polynesian migrations. The method assumes that, if migrants from a single island population settle two different destinations, the variants lost in each migration would be different. So, one of the new settlements might have lost variants A, B, and C, while the other might have lost X, Y, and Z. Then, if the new island population that has lost A, B, and C grows and subsequently sends out migrants to settle another island, the new settlement would still lack the A, B, and C variants but also have lost a new set of variants.

These analyses allowed the authors to reconstruct the migrations that originated in Samoa (around 800 CE) and expanded eastward; they also estimated the times of settlement for the different islands. Some islands appear to serve as hubs for expansions. All of the populations of Eastern Polynesia seem to have been colonized by migrants (either directly or via intermediate islands) from the islands of Rarotonga and Palliser. The most recent settlement (around 1200 CE) was the easternmost island of Rapu Nui (Easter Island), approximately 3,500 kilometers from the coast of Chile.

Given the history of Polynesian exploration and the proximity of Rapu Nui to South America, it's reasonable to wonder whether migrations to or from South America to Polynesia played a role in the settling of the easternmost islands. The Kon-Tiki expedition in 1947, led by the Norwegian explorer and writer Thor Heyerdahl, was a journey by raft to Rapu Nui from South America intended to demonstrate the possibility that Native Americans might have been the original colonizers of Polynesia.[9] Although this hypothesis has been

---

[9] As a teenager, I was fascinated by Heyerdahl's dramatic account of his expedition and many decades, later, had the opportunity to visit the Kon-Tiki museum in Oslo.

rejected by the scientific community in favor of an Asian origin, as we've just seen, the idea of genetic interaction between Polynesians and Native Americans remains an intriguing question. My Norwegian immunogenetics colleague Erik Thorsby searched for Native American HLA alleles in Rapa Nui, in support of his countryman's hypothesis . . . and found some. The critical question was *when* they appeared in the Polynesian community. A recent study of 807 genome sequences from 17 island populations and 15 Pacific Coast Native American groups (Ioannidis et al., 2020, p. 583) provides "evidence for prehistoric contact of Polynesian individuals with Native American individuals (around AD 1200)" in eastern Polynesia before the settlement of Rapu Nui. The Native American sequences found in Eastern Polynesia are most closely related to the indigenous inhabitants of present-day Colombia. The ability to assign time estimates, based on the analysis of these sequences and of the *length* of shared genomic segments and novel statistical methods, allowed the authors to distinguish a contact with Native Americans from Colombia before European colonization from a much later postcolonization contact with Native American groups from Chile. Given the experience and skill of Polynesian seafarers, it seems more likely that Polynesian groups traveled to South American and returned with Native American genes than the Heyerdahl hypothesis of South Americans sailing to Polynesia.

Native Americans and the Peopling of the Americas

In addition to providing insights into the genetic relatedness of different human populations and their migrations, the *pattern* of the allele frequency data can also prove instructive. Some indigenous populations, such as Native Americans, have many fewer HLA alleles than most other populations. (The genetic diversity for other genes is also significantly reduced in Native Americans.) At the other extreme, African populations typically have many more HLA alleles than European or Asian populations. The greater genetic diversity among African population provided critical support for the Out of Africa hypothesis, which proposed that all modern humans in Europe, Asia, the Pacific, and the Americas, arose from groups that migrated out of Africa[10]

---

[10] As noted previously, modern Africans are the descendants of human populations that remained in African and have retained a high degree of genetic diversity.

and colonized the rest of the world, bringing with them only a *subset* of the genetic diversity that was and remains present in Africa.

The dramatic reduction in genetic diversity (i.e., far fewer HLA alleles) in Native American populations compared to other populations and to their presumed Northeast Asian forebears suggests a severe "population bottleneck," a historical event in which a small population capable of containing only a few alleles gave rise to the contemporary Native American population samples analyzed in these genetic studies. One interpretation of the population data is that the reduced diversity observed in contemporary Native American populations reflected the colonization of the Americas some 15,000–25,000 years ago by very small population groups that migrated from Siberia across the Beringian land bridge to North America. The proximity of Native American and Northeast Asian populations in phylogenetic networks based on genetic similarity is consistent with this view, which is also supported by a wealth of physical and cultural anthropological data. An alternative hypothesis for explaining the reduced genetic diversity is a massive population collapse in the Americas due to infections transmitted from the European colonizers.[11] (A reduction in population size would lead to loss of alleles, resulting in reduced diversity.) However, analysis of human remains from pre-Colombian burial grounds revealed the same reduced diversity seen in contemporary Native American populations, supporting the hypothesis that the bottleneck occurred during the migrations from Northeast Asia (Siberia) to the Americas.

As we've just seen in Chapter 10, recent DNA sequencing of ancient indigenous remains as well as modern Native American populations has provided a much more detailed picture with time estimates for the migration from Beringia to North American and subsequently, to South America. These reconstructions of the migrations that led to the peopling of the Americas, based on the genetic analyses of native DNA samples, generated considerable controversy and anger among some of the indigenous Native American populations that had been sampled in these studies.

---

[11] The reduced genetic diversity, particularly of the HLA alleles that influence the immune response, may have rendered these populations more susceptible to novel infectious pathogens.

12

Controversies and Contention
in Interpreting the Past

The origin of Native Americans suggested by population networks or DNA sequence analyses conflicted with many of the creation myths of Native American tribes. There are as many different narratives as there are tribes, but they all have a common theme: *We did not come here from elsewhere; we have always been here.* The genetic analysis of Native American samples, as well as those of other aboriginal groups, became highly contentious, resulting in intense cultural and political controversies. The idea that the indigenous peoples of the Americas might have migrated from elsewhere around 25,000 years ago, an interpretation of data generated by nonnative scientists from native DNA samples, could not be reconciled with their religious traditions and their belief that they were already here at that time and that they had emerged from below the North American earth rather than from Siberia.

Another issue involved informed consent. All studies with human subjects now, and for the last several decades, require that anyone contributing a biological sample, like blood or a cheek swab, for a research project sign a form that describes the nature of the study. This form, which is the ethical bedrock of all population genetics and medical studies, not only memorializes the willingness of the participants to contribute the sample but also indicates that they have understood the aims of the study.[1] The Havasupai, a Native tribe living in the Grand Canyon, sued Arizona State University for facilitating studies that were not covered by the informed consent agreement

---

[1] Violations of this ethical standard are considered very serious. Some of the recent Chinese population genetics studies of the Uyghur minority have been severely criticized as not having met the informed consent requirement amid the more general and well-founded concerns that the Chinese government will use these studies in their well-documented oppression of this minority population. Here, the informed consent violation serves as a kind of "canary in the coal mine" for serious human rights abuses. A recent report by the United Nations human rights office criticized China's mass detention of Uyghers and other Muslim groups in Xinjiang, stating that these actions "may constitute international crimes, in particular crimes against humanity." China is currently collecting DNA samples from Tibetans as part of their mass surveillance program there.

and initiated legal disputes involving many labs. Our lab, which had obtained DNA samples of the Havasupai from a colleague at Arizona State University to analyze HLA allele diversity, was not sued, but we were asked to return all of our samples to the tribe.

I had a long-standing interest in Native American culture and history from my time as a Vista volunteer in New Mexico, and we had obtained samples from several other Native American groups from other sources. Steve Mack and I found patterns of HLA allele diversity that were unique to Native American populations and also consistent with the hypothesis of migration from Siberia to the Americas. Given the controversy surrounding analysis of the Havasupai samples, we returned them all to a legal representative of the Havasupai tribe and decided not to publish any of our findings for this group.

For the collection of the Havasupai samples, the informed consent form specified that the samples would be used for a genetic study of type 2 diabetes, a disease that is unusually common in Native American populations. When Steve and I generated and analyzed the HLA genotyping data, we had been unaware of the restricted nature of the informed consent agreement. The HLA genes are not causally related to type 2 diabetes but, as we and others have shown, are very strongly associated with type 1 diabetes (T1D). A particular HLA genotype confers a 30-fold increased risk. In a study of T1D in Mexican Americans (Erlich et al., 1993), we were able to show that the HLA genetic variants (alleles) that conferred disease susceptibility were European in origin, and that the HLA alleles derived from Native Americans did not confer susceptibility. The incidence of T1D is much higher in European-derived populations than in Native Americans. Mexican American populations have an intermediate incidence, and we suggested that the differences in HLA frequency patterns might account, in part, for this difference in disease incidence. This conclusion was possible only because we had developed a detailed picture of the HLA alleles present in Native American populations.

A problem that was broader and more complicated than the violation of informed consent was the anger directed against the nonindigenous (mostly white) scientists for using the samples collected from aboriginal communities to describe the origins of indigenous populations. As discussed earlier, these historical narratives inferred from the genetic data conflicted with their own creation myths. Each indigenous tribe has its own mythologies, but many of the creation myths include the notion that their ancestors arose from under the earth rather than migrating from Siberia.

In principle, these cultural myths might be able to coexist with scientific data about the genetic relatedness of Native American and Siberian populations, but, in actual human communities, reconciling these disparate narratives is not simple or easy. Many Christians manage to maintain their Genesis-based accounts of human origins in the face of scientific data supporting evolution, but, of course, many do not and continue to try to prevent the teaching of evolution in the public schools.

The Human Genome Diversity Project and Its Discontents

How genetic variation is distributed around the world remains a fascinating and complex scientific issue and one that has a complicated and contentious history. In the 1990s, a concerted international effort to pursue this question, the Human Genetic Diversity Project (HGDP) was initiated, with Luigi Luca Cavalli-Sforza, the pioneering geneticist who first proposed analyzing human migrations with population trees, as its best-known champion. This project proposed to collect and genotype samples of indigenous populations around the world with the goal of analyzing the population genetic frequencies of these isolated populations. The ultimate aim was to document the genetic diversity and reconstruct the history of human migrations that resulted in the world we know today. The hope was to establish a well-funded comprehensive international project with the imprimatur of respected establishment institutions like the National Institutes of Health. However, while significant progress was made by individual labs and in some collaborative studies, this vision of the HGDP was not fully realized due to the objections by some indigenous communities and human rights organizations who perceived issues of scientific racism, colonialism, and informed consent.

Many of these concerns were, based on a long history of colonialist exploitation and oppression, eminently understandable, but I hoped that these issues could be addressed in the meetings discussing the ultimate design of the HGDP. One of the organizations claiming to represent the interests of indigenous communities, the Rural Advancement Foundation International (RAFI)[2] attacked the HGDP, arguing that these moral, ethical, and legal issues had not been adequately addressed. Some of the more incendiary

---

[2] I recently (2020) went to the RAFI website and it describes a benign organization advocating for rural farmers with no hint of its contentious history.

arguments accused the HGDP of stealing and patenting genes from indigenous communities. In testimony to the National Academy of Sciences in 1996, Hope Shand, the research director of RAFI, noted that the HGDP had been "formally opposed in more than 20 declarations issued by indigenous peoples worldwide" and that "UNESCO had declined to endorse the HGDP." She went on. "Given the international controversy and unresolved issues surrounding the HGDP, we believe it would be a grave mistake for the United States government to give further support to it."

Critics of the concept of "pure" research argue that politics and issues of power cannot be separated from science. In this case, the politics, reflecting the mistrust by the Global South of the Global North, ended up curtailing scientific inquiry as well as damaging the careers of some scientific colleagues. My own lab and reputation became embroiled in this contentious political narrative, as I had participated in some of the early discussions about the idea of establishing an HGDP. At one of these meetings, I met an English documentary filmmaker, Luke Holland, who was interested in making a documentary about this kind of genetics research. I put Holland in touch with a close friend and colleague, Tom White (Cetus and Roche Molecular Systems [RMS]). Tom had been in discussions with two Colombian scientists, Drs. Genoveva Keyeux and Jaime Bernal of the Universidad Javeriana in Bogota about an ambitious project, La Gran Expedicion Humana, to visit all of the indigenous communities in Colombia, to provide medical and dental treatment and take cheek swab samples for genetic analysis of the heritable diseases that affected some of these tribal populations. Tom and I sent them the material to carry out some of the PCR-based genetic typing we had developed, and we offered to apply our HLA genotyping system to the samples they had collected. Tom and one of my postdoctoral fellows, Beth Trachtenberg, accompanied the Colombian doctors and scientists on one of these expeditions to the Putumayo River region in 1993. They both told me it had been a truly thrilling and fascinating experience, and I deeply regretted that I had been too busy at the time to join the team. Luke Holland and his filmmaker partner, Ian Taylor, decided to fund another trip to visit these communities with two of my other postdoctoral fellows so that the filmmakers could accompany the expedition and film the whole experience. No samples were collected in this second trip, funded by and basically "staged" for the filmmakers. Although Tom and I were employed by RMS, a company that produced PCR-based diagnostic tests, our involvement was based solely on our personal basic research interests in human population genetics and Tom's friendship with the Colombian scientists. RMS had no material interest or involvement with any part of this project.

Unfortunately, none of this mattered to Holland and Taylor. They produced a documentary, *The Gene Hunters* (1996), that accused the HGDP and La Gran Expedicion Humana, as well as Roche, the parent company of RMS, of carrying out a project to exploit local indigenous tribes, patenting their genes for profit. Tom and I, as well as the postdoctoral fellows who were filmed accompanying the team, were portrayed as representatives of a multinational company whose goal was commercial exploitation. We tried repeatedly to correct this misrepresentation, as did Professor Bernal, but Holland and Taylor preferred their false narrative. I don't know if they didn't care or if they found it simply inconceivable that Roche, a multinational corporation, was not ultimately involved in this project. A narrative that fits preconceived notions, even one based on a lie, can take on a life of its own, and the filmmakers were determined not to let the facts get in the way of a good story. The fallout from this controversy was significant. Dr. Geneva Keyeux, a dedicated scientist, lost her position at the university and the reputation of La Gran Expedicion Humana, a project that brought invaluable medical and dental treatment to isolated tribal communities was damaged beyond repair.

The systematic analysis of DNA sequence and population frequency data proposed by Cavalli-Sforza (the HGDP) was never established as the large international project he envisioned. The sensitivities of the organizers to the concerns of indigenous populations were, arguably, insufficient. Nonetheless, the data on contemporary populations from individual labs and consortia, working independently of the HGDP, have transformed our sense of who we are and where we've come from. These data have provided a rich and detailed picture of human genetic diversity and human migrations, as illustrated in Figure 11.1 in Chapter 11. One critical source of genetic data illuminating our history was not even imaginable by Cavalli-Sforza and his HGDP colleagues. DNA sequence analysis of archaic human subspecies, such as Neanderthals and Denisovans, as well as of ancient human remains, as we've seen in Chapter 10, has provided a new and immensely powerful perspective on human migrations.

Very recently (2022), working with novel DNA sequencing technology, more sophisticated statistical methods, and a better appreciation of the need to work with indigenous communities, a new genome diversity project has been initiated, with the blessing (i.e., funding) of the US National Human Genome Research Institute and the National Institutes of Health around

30 years after the initial HGDP discussions. The goal of this international consortium of scientists is to generate a reference genome (The Human Pangenome Project) that captures the genetic diversity of all the people of the world (Wang et al., 2022). Currently, the reference human genome used by geneticists around the world is a composite of 3.1 billion bases, made from around 20 individuals, with most of the sequence derived from a single individual. The initial goal of the project is to generate complete genome sequences from 350 different individuals from diverse populations, including indigenous groups, after obtaining consent and collaboration. Eventually, the project organizers hope to sequence thousands of genomes to capture even more of the genetic diversity present in modern human populations.

13

Ancestry and Genetics

What's Race Got to Do with It?

The allele frequency databases for ethnically defined populations that help reconstruct human migrations and help interpret forensic matches are now also being used extensively to explore the genetic heritage of individuals. Trying to discover our individual ancestry through genetic analysis is now a popular pastime (sometimes known as "recreational genetics") and the basis for the commercial success of consumer genetic testing companies like 23andMe and Ancestry.com.[1] The first direct-to-consumer ancestry testing company, Oxford Ancestors,[2] which assigned clients to one of The Seven Daughters of Eve, based on mitochondrial DNA, provided, arguably, more entertainment than meaningful information but, inarguably, stimulated the public interest in genetics and ancestry. Although the details of individual genetic composition (43% This, 23% That, and 34% The Other) can vary in the reports from different companies using different databases and different algorithms, these analyses of cheek swabs or saliva have provided millions of people personally meaningful insights into their genetic heritage.

These genetic genealogy databases are, of course, also being used for other purposes. As we discussed in Part I, many long-standing cold cases, such as the Golden State Killer, have been solved by "familial searching" by law enforcement of some genealogy databases. Given significant concerns about privacy, this application of genealogy databases now requires informed consent from the participants, and several states (e.g., Maryland) have passed legislation providing stringent guidelines for these investigative searches (discussed in Chapter 6). In addition, some companies, such as 23andMe are providing the databases to pharmaceutical companies to aid in their search for new therapies and diagnostic tests. So the Silicon Valley cliché, "if you're

---

[1] The public offering for 23 and Me valued the company at $3.5 billion, and the Blackstone Group acquired a majority stake in Ancestry.com for $4.7 billion.
[2] Oxford Ancestors was founded by Brian Sykes, a professor of human genetics at Oxford University. He passed away in 2020. His company no longer exists.

not paying for it, then you are the product" needs revision in light of this practice. Even if you *are* paying (a modest amount, to be sure), you may still be the product.

These ancestry companies provide an ancestry composition based on comparisons with various ethnic population databases. Another category of statistical programs (e.g., STRUCTURE, developed by Jonathan Pritchard, and FRAPPE, developed by Hua Tang) analyzes a collection of *individual* datapoints and provides an estimate of ancestry composition without reference to population databases. This software does not attempt to plot the genetic relatedness of the individual samples, as principal component analysis (PCA) does, but gives a breakdown into the ethnic components of each individual sample (Figure 13.1). PCA, as we've seen in Chapter 11, can array individual genotypes in a two-dimensional pattern based on genetic relatedness. STRUCTURE and FRAPPE also analyze individual genotypes without reference to predefined groups, but these software programs display the ancestral genetic diversity *within* each individual sample. A unique feature of these two programs is that the number of different ethnic components can be defined and the composition analysis of each individual sample is not based on reference population databases but on the overall sample collection. Figure 13.1 K = 3 illustrates the ancestral composition analysis of various Hispanic individual genotypes based on *three* different ancestral groups

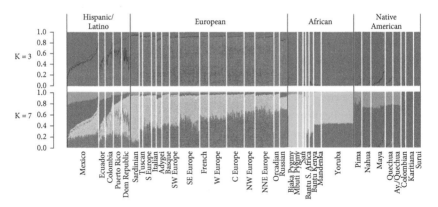

Figure 13.1   Frappe clustering illustrating the admixed ancestry of Hispanic/ Latinos shown for K = 3 and K = 7. Individuals are shown as vertical bars colored in proportion to their estimated ancestry within each cluster. Native American populations are listed in order geographically, from north to south (from Bryc et al., 2010).

ANCESTRY AND GENETICS   179

while the analysis in Figure 13.1 K = 7 is based on *seven* different groups. The software feature that can specify any number of different groups illustrates the fluidity of our concepts of discrete ancestral categories.

These two software programs provide valuable data for academic researchers, but it is the ancestry analyses provided by the commercial genealogy companies that have turned our fascination with genomics and ancestry into big business.[3] However, the concepts we use to describe population differences and genetic heritage—ancestry, ethnicity, and most problematically, race—are all complicated and inextricably intertwined with social factors like language, religion, and cultural traditions.[4] One way to think about the relationship of genes and culture in ancestry and ethnic identity is to consider the transmission of chromosomes and genes from parents to two offspring. For any gene, the probability that each child would inherit the *same* two alleles from their parents is 25% and the probability that they would share *zero* alleles is also 25%.[5] So there is a significant probability that two siblings who share the same cultural identity might have inherited a significantly different genetic legacy. Similarly, if I had two second-degree cousins, I might share a very different amount of DNA with them but my familial (ancestral) relationship to them would be the same.

As we've seen in the previous pages and, more importantly, in the world around us, discussions about the relationship of genetics to population differences, given the tragic history of racism in our society, have always been complicated, contentious, and too often confused. Of all the words we use to characterize population differences, "race," "ancestry," "ethnicity," it is the concept of "race" that has done the most damage throughout human history. By now, it is clear to almost all scientists and most informed members of the public that the traditional notion of "race" is a social construct without a well-defined biological basis. Franz Boas, the father of modern cultural anthropology, was, in the early 1900s, one of the first scholars to oppose the then popular ideology of "scientific racism," the idea that (1) humans could be categorized into a few (4 or 5) discrete groups ("races"), (2) these groups

---

[3] The Harvard geneticist George Church, founder of Nebula Genomics, has announced he is auctioning off his genome sequence as an NFT (nonfungible token).

[4] A thoughtful discussion and definition of these terms is presented in Peterson et al. (2019).

[5] The probability that the two kids inherited the same allele from one parent and different alleles from the other parent is 50%. The biological mechanism underlying this pattern of inheritance is the segregation of chromosomes during meiosis. The principle of independent assortment of traits was discovered by Gregor Mendel in his study of pea shape and color. The discovery that the independent assortment of traits (eye color in fruit flies) was mediated by chromosome segregation was made by Thomas Hunt Morgan in 1910.

exhibited behavioral differences, (3) these groups could be defined by biology and are genetically homogeneous, and (4) one group (Europeans, in particular, northern Europeans) was superior to the other groups. That racial and ethnic categories are socially constructed is clear from the notorious "one-drop" rule for defining African heritage[6] in the United states and is elegantly captured in the title of Noel Ignatiev's 1996 book *How the Irish Became White*.[7] Tribalism (Us vs. Them) and the claims of superiority of one group over another have, sadly, been part of human culture for all of recorded history. However, the tenets of "scientific racism" in the early 19th and 20th centuries, used to justify slavery, colonialism, and restrictive immigration policies, were particularly toxic in that they claimed to be supported by scientific evidence. Tragically, the belief in biologically determined behavioral differences continues in white supremacist groups in the United States and around the world.

However, as we've seen in the preceding pages, there *are* some genetic differences between populations with different biogeographic ancestries. The analysis with the genetic markers of ancestry (ancestry informative markers, or AIMS) or the clustering analysis shown in the PCA in Figure 11.5 focus on the tiny subset of genetic variants whose frequency *does* differ in different population groups. As many geneticists have pointed out, (and as I discussed in Part I), well over 99% of our genome is identical in all human populations. (In this context, it's worth remembering that the genomes of chimpanzees and humans are over 98% identical.) In my view, the most important observation about genetics and human populations was made many years ago by the population geneticist Richard Lewontin,[8] who pointed out that there was more genetic variation *within* populations than *between* populations. This fundamental insight into how genetic differences are distributed among human populations is all the more remarkable in that it was made based on the analysis of a few proteins, well before the development of the DNA technology that has transformed our current understanding of human genetics. The analysis of DNA sequences from hundreds of thousands of human

---

[6] This social and legal principle, codified into law in some states in the 20th century, asserts that someone with any black ancestor ("one drop of black blood") is classified as "black."

[7] Another example of the arbitrary and ill-defined concept of "race" and its relationship to skin color is the recent brouhaha about Whoopi Goldberg's statement that the Holocaust was not about race; it was just some white people killing other white people.

[8] Lewontin was one of the geneticists who, with his colleague Dan Hartl, argued forcefully against the early use of DNA analysis in forensics, as discussed in Chapter 2 of Part I. In his chosen role as an eloquent contrarian, he also was a critic, misguided in my view, of the Human Genome Project. He passed away in 2021.

samples has confirmed Lewontin's initial and immensely consequential observation.

What does this characterization of how genetic variation is distributed within and between populations actually mean? One simple, arguably simplistic, way to think about this is to consider two populations on different continents: Population A has 10 different variants (variant #1–variant #10) at a particular gene while population B has 10 variants (variant #2–variant#11) all at the same frequency, so that variants #2–#10 are shared by both populations. There is significant variation within each population (10 different variants) but most of the variation is shared between the two populations. However, variants #1 and #11 could serve as the AIMS, discussed earlier. Conceivably, variants #1 and #11 could have arisen after the populations separated, or both variants might have been present in the ancestral population but experienced changes in their frequencies (dropping to zero in one of the two populations) due to random chance (genetic drift). Whatever the origins of the current variation patterns, we can choose to focus either on the small genetic differences that distinguish populations or on the variation within each population that is shared. By focusing on the genetic variation that is associated with biogeographical ancestry, it is possible using various software to (1) cluster individuals into groups corresponding to geography (PCA) or (2) assign different proportions of geographic ancestry to individuals (STRUCTURE, FRAPPE, or the algorithms used by commercial ancestry companies). The association of these selected variants with ancestry should not obscure the fact that the vast majority of human genetic variation is among individuals within a given population rather than between populations.

One area where the concept of race can, arguably, be useful, as a "convenient fiction," is medicine. Some genetic variants associated with specific diseases are more common in some populations. A classic example, the beta-globin mutation that causes sickle cell anemia,[9] an autosomal recessive disease (requires *two* copies of the mutation), is more common in West Africa and in people with West African ancestry. Carriers (people with *one* copy of the mutation but no disease) are resistant to malaria. This selective advantage of the carrier is the cause of the increased frequency of the mutation in some populations where malaria has been endemic. The sickle mutation in

---

[9] As noted in Part I, when we were developing PCR, the first gene we amplified was the beta-globin gene for the diagnosis of sickle cell anemia.

beta-globin can also be found in parts of the Mediterranean and in Southeast Asia, but its frequency in highest in sub-Saharan Africa. Other mutations or genetic variants conferring susceptibility to specific diseases are more frequent in some population groups than in others. For managing individual patients, it would, of course, be more informative to genotype the patient ("personalized medicine") than to ask about membership in some ill-defined group. At the moment, while genome sequencing would be preferable, ancestry still has some value as a crude proxy for genetic information.

In thinking about "race," we can acknowledge that there are some genetic differences between populations while still fiercely opposing the toxic idea that some behavioral patterns can be attributed to these differences and also that some groups are "superior" to others. We scientists have a particular responsibility to address these issues because it was the work of some early geneticists[10] that was invoked as the scientific basis of these racist beliefs.

As we've seen in Part I, over the past three decades, DNA analysis has played a critical role in making the criminal justice system more reliable and more just. We can hope that the analysis of the genetic differences between individuals and populations, both living and extinct, discussed in Part II, can lead to a deeper understanding of our complex history and an appreciation of our common genetic heritage.

[10] The British scientist Francis Galton, the father of human genetics, who coined the term "eugenics," claimed the genetic superiority of the white English upper classes in his 1869 *Hereditary Genius*.

Appendix

Polymerase Chain Reaction (PCR)

PCR is a technology for amplifying DNA, creating millions of copies of a specific targeted DNA sequence in a test tube, using the same strategy as DNA replication within the cell. PCR consists of a series of cycles, each of which has three steps. In the first step, the two strands of a double-stranded DNA molecule are separated so that each strand can serve as a template for the synthesis of a new strand in subsequent steps. This "denaturation" (separation) of the double-stranded DNA molecule is achieved by heating the reaction to a high temperature (around 95°C). In the second step, short synthetic pieces of DNA, known as PCR primers, designed to be complementary to specific sequences in the targeted DNA, bind ("anneal") to the separated strands. These primers have a chemical polarity such that one end of each primer, the 3' end, can be extended, leading to synthesis of new DNA strands. In this second step, the primers flank the targeted region, bind to opposite strands, and are oriented with their 3' ends facing each other. The primers bind to the DNA at a lower temperature (around 60°C). In the third step, an enzyme, DNA polymerase, synthesizes two new DNA strands, extending the primers on their respective template strand, doubling the amount of the targeted DNA. With the currently available set of thermostable (resistant to the high temperature used in the first step) DNA polymerases, this step is typically carried out at around 70°C. Every cycle doubles the amount of the targeted DNA, so that this exponential reaction can produce a million copies in 20 cycles.

A Schematic Diagram of PCR

Many modifications of this basic PCR strategy have been developed since our first publication in 1985. One of the most critical changes was the replacement of the *E. coli* DNA polymerase (Klenow), which was inactivated at high temperature, such that new DNA polymerase had to be added to the reaction after every cycle. The first thermostable DNA polymerase used in PCR was Taq polymerase from the thermophilic bacterium *Thermus aquaticus*. Now there are many available. Incorporating a thermostable DNA polymerase into PCR meant that the critical components of a PCR (DNA, primers, DNA polymerase, the nucleotide building blocks for DNA synthesis, and the appropriate concentration of cofactors required for enzyme activity) could all be assembled in a test tube and PCR amplification could be achieved by simply cycling the temperature. In fact, the instruments designed to automate PCR were known as "thermal cyclers."

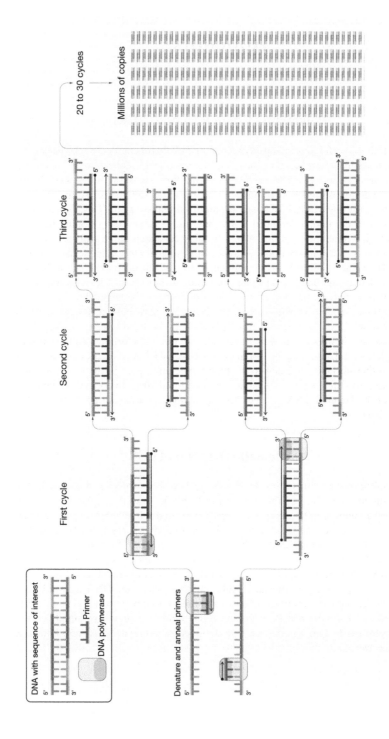

DNA with sequence of interest

Primer

DNA polymerase

Denature and anneal primers

First cycle

Second cycle

Third cycle

20 to 30 cycles

Millions of copies

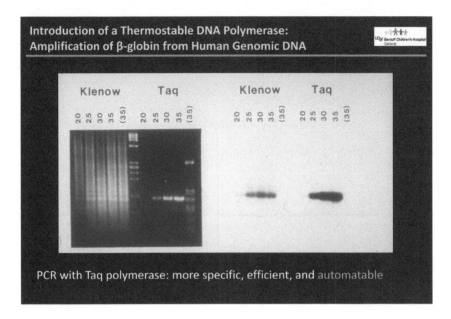

Analysis of PCR amplification products using *E. coli* DNA polymerase (Klenow) or Taq polymerase

On the left, the PCR amplification with primers for the beta-globin gene is shown from 20–35 PCR cycles with the *E.coli* DNA polymerase (Klenow) and the thermostable Taq polymerase. The PCR products, generated by varying numbers of PCR cycles, are analyzed by gel electrophoresis. The lane indicated by (35) is the PCR amplification for 35 cycles of a human genomic DNA with the beta-globin DNA deleted to serve as a negative control. The PCR products are stained with a dye so that *all* of the amplified DNA is detected. On the right, the PCR products in the gel are analyzed with a radioactive probe specific for beta-globin so that *only* the amplified beta-globin DNA is detected. This procedure is known as a Southern blot. Comparing the left and right results for PCR amplification using the *E. coli* Klenow polymerase indicates that, with a specific probe, the amplification of a defined beta-globin DNA fragment of the expected size can be detected. *In the absence of the probe, there is no evidence of beta-globin amplification.* In fact, additional experiments revealed that only 1% of the DNA amplified with the *E. coli* DNA polymerase was beta-globin. Had we not used a specific probe, we would have been unable to demonstrate that the concept of PCR actually "worked." With the probe, we could show a 100,000-fold enrichment of the targeted beta-globin gene (Saiki et al., 1985). The incorporation of the thermostable DNA polymerase resulted in a more specific and efficient amplification (Saiki et al., 1988). Even without the beta-globin probe, a defined DNA fragment of the expected size can be detected.

References

Ayala, Francisco J. "The Myth of Eve: Molecular Biology and Human Origins: F. J. Ayala." *Science* 270, no. 5244 (1995): 1930–36.

Bali, Gunmeet Kaur, Mary Wisner, Steven J. Mack, Sandy Calloway, and Henry Erlich. "Next Generation Sequencing using Probe Capture: De-convolution of Mixtures with Mitochondrial DNA, STR, and SNP Markers from a Single Shotgun DNA Library." Abstract. California Association of Criminalists (2021).

Banuelos, M. M., Y. J. A. Zavleta, A. Roldan, R. J. Reyes, M. Guardado, M. Chavez, et al. "Associations between Forensic Loci and Expression Levels of Neighboring Genes may Compromise Medical Privacy." *Proceeding of the National Academy Sciences U.S.A.* 119, no. 40 (2022).

Bergström, Tomas F., Agnetha Josefsson, Henry A. Erlich, and Ulf Gyllensten. "Recent Origin of HLA-DRB1 Alleles Bieber, Brenner, and Lazar 2006 and Implications for Human Evolution." *Nature Genetics* 18, no. 3 (1998): 237–42.

Bieber, Frederick R., Charles H. Brenner, and David Lazer. "Finding Criminals through DNA of Their Relatives." *Science* 312, no. 5778 (2006): 1315–16.

Blake, Edward, Jennifer Mihalovich, Russell Higuchi, P. Sean Walsh, and Henry Erlich. "Polymerase Chain Reaction (PCR) Amplification and Human Leukocyte Antigen (HLA)-DQα Oligonucleotide Typing on Biological Evidence Samples: Casework Experience." *Journal of Forensic Sciences* 37, no. 3 (1992): 700–26.

Boehnke, M., N. Arnheim, H. Li, and F.S. Collins. "Fine-structure genetic mapping of human chromosomes using the polymerase chain reaction on single sperm: experimental design considerations." *American Journal of Human Genetics* 45 (1989): 21–32.

Brace, Selina, Yoan Diekmann, Thomas Booth, Ruairidh Macleod, Adrian Timpson, Will Stephen, Giles Emery, Sophie Cabot, Mark G. Thomas, and Ian Barnes. "Genomes from a Medieval Mass Burial Show Ashkenazi-Associated Hereditary Diseases Pre-Date the 12th Century." *Current Biology* 32, no. 20 (2022). Epub.

Bright, Jo-Anne, Kevin Cheng, Zane Kerr, Catherine McGovern, Hannah Kelly, Tamyra R. Moretti, Michael A. Smith, et al. "Strmix™ Collaborative Exercise on DNA Mixture Interpretation." *Forensic Science International: Genetics* 40 (2019): 1–8.

Bright, Jo-Anne, Duncan Taylor, Catherine McGovern, Stuart Cooper, Laura Russell, Damien Abarno, and John Buckleton. "Developmental Validation of Strmix™, Expert Software for the Interpretation of Forensic DNA Profiles." *Forensic Science International: Genetics* 23 (2016): 226–39.

Bryc, Katarzyna, Christopher Velez, Tatiana Karafet, Andres Moreno-Estrada, Andy Reynolds, Adam Auton, Michael Hammer, Carlos D. Bustamante, and Harry Ostrer. "Genome-Wide Patterns of Population Structure and Admixture among Hispanic/ Latino Populations." *Proceedings of the National Academy of Sciences* 107, no. supplement 2 (2010): 8954–61.

Butler, John M., Margaret C. Kline, and Michael D. Coble. "NIST Interlaboratory Studies Involving DNA Mixtures (mix05 and mix13): Variation Observed and Lessons Learned." *Forensic Science International: Genetics* 325, no. 6099 (2018): 81–94.

Cann, Rebecca L., Mark Stoneking, and Allan C. Wilson. "Mitochondrial DNA and Human Evolution." *Nature* 325, no. 6099 (1987): 31–36.

Cavalli-Sforza, L. L., P. Menozzi, and A. Piazza. *The History and Geography of Human Genes*. Princeton, NJ: Princeton University Press, 1994.

Chakraborty, Ranajit, and Kenneth K. Kidd. "The Utility of DNA Typing in Forensic Work." *Science* 254, no. 5039 (1991): 1735–39.

Chamberlain, Michael. "Familial DNA Searching: A Proponent's Perspective." *Criminal Justice* 27 (2012): 18.

Chen, Fahu, Frido Welker, Chuan-Chou Shen, Shara E. Bailey, Inga Bergmann, Simon Davis, Huan Xia, et al. "A Late Middle Pleistocene Denisovan Mandible from the Tibetan Plateau." *Nature* 569, no. 7756 (2019): 409–12.

Coble, Michael D., Odile M. Loreille, Mark J. Wadhams, Suni M. Edson, Kerry Maynard, Carna E. Meyer, Harald Niederstätter, et al. "Mystery Solved: The Identification of the Two Missing Romanov Children Using DNA Analysis." *PLoS ONE* 4, no. 3 (2009): e4838.

Erlich, Henry A., Christian López-Peña, Katie T. Carlberg, Shelly Shih, Gunmeet Bali, Ken D. Yamaguchi, Hugh Salamon, Reena Das, Ashutosh Lal, and Cassandra D. Calloway. "Noninvasive Prenatal Test for β-Thalassemia and Sickle Cell Disease Using Probe Capture Enrichment and Next-Generation Sequencing of DNA in Maternal Plasma." *Journal of Applied Laboratory Medicine* 7, no. 2 (2021): 515–31.

Erlich, Henry A., Eric Stover, Thomas J. White, and Scott Turow. *Silent Witness: Forensic DNA Evidence in Criminal Investigations and Humanitarian Disasters*. New York: Oxford University Press, 2020.

Erlich, Henry A., Adina Zeidler, Julie Chang, Sylvia Shaw, Leslie J. Raffel, William Klitz, Yolanda Beshkov, et al. "HLA Class II Alleles and Susceptibility and Resistance to Insulin Dependent Diabetes Mellitus in Mexican-American Families." *Nature Genetics* 3, no. 4 (1993): 358–64.

Gill, Peter, and Erika Hagelberg. "Ongoing Controversy over Romanov Remains." *Science* 306, no. 5695 (2004): 407–10.

Gill, Peter, Pavel L. Ivanov, Colin Kimpton, Romelle Piercy, Nicola Benson, Gillian Tully, Ian Evett, Erika Hagelberg, and Kevin Sullivan. "Identification of the Remains of the Romanov Family by DNA Analysis." *Nature Genetics* 6, no. 2 (1994): 130–35.

Gill, Peter, Alec J. Jeffreys, and David J. Werrett. "Forensic Application of DNA 'Fingerprints.'" *Nature* 318, no. 6046 (1985): 577–79.

Green, Richard E., Anna-Sapfo Malaspinas, Johannes Krause, Adrian W. Briggs, Philip L. F. Johnson, Caroline Uhler, Matthias Meyer, et al. "A Complete Neandertal Mitochondrial Genome Sequence Determined by High-Throughput Sequencing." *Cell* 134, no. 3 (2008): 416–26.

Halldorsson, B.V., H.P. Eggertsson, K.H.S. Moore, H. Hauswedell, O. Eiriksson, et al. "The sequences of 150, 119 genomes in the U.K. Biobank." *Nature* 607 (2022): 732–40.

Higuchi, Russell, Barbara Bowman, Mary Freiberger, Oliver A. Ryder, and Allan C. Wilson. "DNA Sequences from the Quagga, an Extinct Member of the Horse Family." *Nature* 312, no. 5991 (1984): 282–84.

Higuchi, Russell, Cecilia H. von Beroldingen, George F. Sensabaugh, and Henry A. Erlich. "DNA Typing from Single Hairs." *Nature* 332, no. 6164 (1988): 543–46.

Hofreiter, Michael, Odile Loreille, Deborah Ferriola, and Thomas J. Parsons. "Ongoing Controversy over Romanov Remains." *Science* 306, no. 5695 (2004): 407–10.

Ingram, M., H. Kaessmann, S. Paabo, and U. Gyllensten. "Mitochondrial Genome Variation and the Origin of Modern Humans." *Nature* 408 (2000): 708–13.

Ioannidis, Alexander G., Javier Blanco-Portillo, Karla Sandoval, Erika Hagelberg, Carmina Barberena-Jonas, Adrian V. Hill, Juan Esteban Rodríguez-Rodríguez, et al. "Paths and Timings of the Peopling of Polynesia Inferred from Genomic Networks." *Nature* 597, no. 7877 (2021): 522–26.

Ioannidis, Alexander G., Javier Blanco-Portillo, Karla Sandoval, Erika Hagelberg, Juan Francisco Miquel-Poblete, J. Víctor Moreno-Mayar, Juan Esteban Rodríguez-Rodríguez, et al. "Native American Gene Flow into Polynesia Predating Easter Island Settlement." *Nature* 583, no. 7817 (2020): 572–77.

Jeffreys, Alec J., John F. Brookfield, and Robert Semeonoff. "Positive Identification of an Immigration Test-Case Using Human DNA Fingerprints." *Nature* 317, no. 6040 (1985): 818–19.

Jeffreys, Alec J., Victoria Wilson, and Swee Lay Thein. "Hypervariable 'Minisatellite' Regions in Human DNA." *Nature* 314, no. 6006 (1985a): 67–73.

Jeffreys, Alec J., Victoria Wilson, and Swee Lay Thein. "Individual-Specific 'Fingerprints' of Human DNA." *Nature* 316, no. 6023 (1985b): 76–79.

Jonsson, Hakon, Erna Magnusdottir, Hannes P. Eggertsson, Olafur A. Stefansson, Arnadottir A. Gudny, Ogmundur Eiriksson, Florian Zink, et al. "Differences between Germline Genomes of Monozygotic Twins." *Nature Genetics* (2021): 53, 27–34.

King, Turi E., Gloria Gonzalez Fortes, Patricia Balaresque, Mark G. Thomas, David Balding, Pierpaolo Maisano Delser, Rita Neumann, et al. "Identification of the Remains of King Richard III." *Nature Communications* 5, No. 5361 (2014).

Kittler, Ralf, Manfred Kayser, and Mark Stoneking. "Molecular Evolution of Pediculus Humanus and the Origin of Clothing." *Current Biology* 13, no. 16 (2003): 1414–17.

Kjær, Kurt H., Mikkel Winther Pedersen, Bianca De Sanctis, Binia De Cahsan, Thorfinn S. Korneliussen, Christian S. Michelsen, Karina K. Sand, et al. "A 2-Million-Year-Old Ecosystem in Greenland Uncovered by Environmental DNA." *Nature* 612, no. 7939 (2022): 283–91.

Knight, A., L. A. Zhivotovsky, D. H. Kass, D. E. Litwin, L. D. Green, P. S. White, and J. L. Mountain. "Molecular, Forensic and Haplotypic Inconsistencies Regarding the Identity of the Ekaterinburg Remains." *Annals of Human Biology* 31, no. 2 (2004): 129–38.

Lander, Eric S. "DNA Fingerprinting on Trial." *Nature* 339, no. 6225 (1989): 501–5.

Lander, Eric S., and Bruce Budowle. "DNA Fingerprinting Dispute Laid to Rest." *Nature* 371, no. 6500 (1994): 735–38.

Levenson, Michael. "Brothers, Wrongfully Convicted of Murder, Are Freed after 25 Years in Prison." *New York Times*, March 24, 2022. https://www.nytimes.com/2022/03/23/us/michigan-brothers-exonerated-murder.html.

Lewontin, Richard. *It Ain't Necessarily So: The Dream of the Human Genome and Other Illusions*. New York: New York Review of Books, 2001.

Lewontin, Richard C., and Daniel L. Hartl. "Population Genetics in Forensic DNA Typing." *Science* 254, no. 5039 (1991): 1745–50.

Lindahl, Tomas. "Instability and Decay of the Primary Structure of DNA." *Nature* 362, no. 6422 (1993): 709–15.

Mack, S. J., T. L. Bugawan, P. V. Moonsamy, J. A. Erlich, H. Trachtenberg, Y. K. Paik, A. B. Begovich, et al. "Evolution of Pacific/Asian Populations Inferred from HLA Class II Allele Frequency Distributions." *Tissue Antigens* 55, no. 5 (2000): 383–400.

Mack, S., and H. Erlich. "13th International Histocompatibility Workshop Anthropology/HumanGenetic Diversity Joint Report—Chapter 6: Population Relationships as Inferred from Classical HLA Genes." In *Immunobiology of the Human MHC:*

Proceedings of the 13th International Histocompatibility Workshop and Conference, edited by J. Hansen, 747–57. Seattle: IHWG Press; 2006.

Marshall, Charla, Kimberly Sturk-Andreaggi, Erin M. Gorden, Jennifer Daniels-Higginbotham, Sidney Gaston Sanchez, Željana Bašić, Ivana Kružić, et al. "A Forensic Genomics Approach for the Identification of Sister Marija Crucifiksa Kozulić." *Genes* 11, no. 8 (2020): 938.

McNamara, Michelle. *I'll Be Gone in the Dark: One Woman's Obsessive Search for the Golden State Killer*. New York: HarperCollins, 2018.

Moreno-Mayar, J. Víctor, Ben A. Potter, Lasse Vinner, Matthias Steinrücken, Simon Rasmussen, Jonathan Terhorst, John A. Kamm, et al. "Terminal Pleistocene Alaskan Genome Reveals First Founding Population of Native Americans." *Nature* 553, no. 7687 (2018): 203–7.

Pääbo, S., R. G. Higuchi, and A. C. Wilson. "Ancient DNA and the Polymerase Chain Reaction." *Journal of Biological Chemistry* 264, no. 17 (1989): 9709–12.

Perlin M., J. M. Hornyak, G. Sugimoto, and K. M. J. Miller. "TrueAllele Genotyping Identificaation on DNA Mixtures Containing Up to Five Unkown Contributors." *Journal of Forenwci Science* 4 (2015): 857–68.

Peñalver, Enrique, Antonio Arillo, Xavier Delclòs, David Peris, David A. Grimaldi, Scott R. Anderson, Paul C. Nascimbene, and Ricardo Pérez-de la Fuente. "Ticks Parasitised Feathered Dinosaurs as Revealed by Cretaceous Amber Assemblages." *Nature Communications* 8, no. 1 (2017): 1924–27.

Peterson, Roseann E., Karoline Kuchenbaecker, Raymond K. Walters, Chia-Yen Chen, Alice B. Popejoy, Sathish Periyasamy, Max Lam, et al. "Genome-Wide Association Studies in Ancestrally Diverse Populations: Opportunities, Methods, Pitfalls, and Recommendations." *Cell* 179, no. 3 (2019): 589–603.

Pinson, Anneline, Lei Xing, Takashi Namba, Nereo Kalebic, Jula Peters, Christina Eugster Oegema, Sofia Traikov, et al. "Human TKTL1 Implies Greater Neurogenesis in Frontal Neocortex of Modern Humans than Neanderthals." *Science* 377, no. 6611 (2022). Epub.

Pollen, Alex A., Umut Kilik, Craig B. Lowe, and J. Gray Camp. "Human-Specific Genetics: New Tools to Explore the Molecular and Cellular Basis of Human Evolution." *Nature Reviews Genetics* 10 (2023): 687–711.

Posth, Cosimo, He Yu, Ayshin Ghalichi, Hélène Rougier, Isabelle Crevecoeur, Yilei Huang, Harald Ringbauer, et al. "Palaeogenomics of Upper Palaeolithic to Neolithic European Hunter-Gatherers." *Nature* 615, no. 7950 (2023): 117–26.

Reich, David. *Who We Are and How We Got Here: Ancient DNA and the New Science of the Human Past*. New York: Pantheon Books, 2018.

Reich, David, Richard E. Green, Martin Kircher, Johannes Krause, Nick Patterson, Eric Y. Durand, Bence Viola, et al. "Genetic History of an Archaic Hominin Group from Denisova Cave in Siberia." *Nature* 468, no. 7327 (2010): 1053–60.

Saiki, R. K., D. H. Gelfand, S. Stoffel, S. J. Scharf, R. Higuchi, G. T. Horn, K. B. Mullis, and H. A. Erlich. "Primer-Directed Enzymatic Amplification of DNA with a Thermostable DNA Polymerase." *Science* 239, no. 4839 (1988): 487–91.

Saiki, Randall K., Stephen Scharf, Fred Faloona, Kary B. Mullis, Glenn T. Horn, Henry A. Erlich, and Norman Arnheim. "Enzymatic Amplification of β-Globin Genomic Sequences and Restriction Site Analysis for Diagnosis of Sickle Cell Anemia." *Science* 230, no. 4732 (1985): 1350–54.

Santos, Fernanda. "DNA Evidence Frees a Man Imprisoned for Half His Life." *New York Times*, September 21, 2006. https://www.nytimes.com/2006/09/21/nyregion/21dna.html.

Severson, Alissa L., Brian F. Byrd, Elizabeth K. Mallott, Amanda C. Owings, Michael DeGiorgio, Alida de Flamingh, Charlene Nijmeh, et al. "Ancient and Modern Genomics of the Ohlone Indigenous Population of California." *Proceedings of the National Academy of Sciences* 119, no. 13 (2022). Epub.

Skov, Laurits, Stéphane Peyrégne, Divyaratan Popli, Leonardo N. Iasi, Thibaut Devièse, Viviane Slon, Elena I. Zavala, et al. "Genetic Insights into the Social Organization of Neanderthals." *Nature* 610, no. 7932 (2022): 519–25.

Slon, Viviane, Fabrizio Mafessoni, Benjamin Vernot, Cesare de Filippo, Steffi Grote, Bence Viola, Mateja Hajdinjak, et al. "The Genome of the Offspring of a Neanderthal Mother and a Denisovan Father." *Nature* 561, no. 7721 (2018): 113–16.

Stoneking, M., D. Hedgecock, R. G. Higuchi, L. Vigilant, and H. A. Erlich. "Population Variation of Human mtDNA Control Region Sequences Detected by Enzymatic Amplification and Sequence-Specific Oligonucleotide Probes." *American Journal of Humane Genetics* 48, no. 2 (1991): 370–82.

Taylor, Duncan, Jo-Anne Bright, and John Buckelton. "The Interpretation of Single Source and Mixed DNA Profiles." *Journal of Forensic Sciences International Genetics* 5 (2013): 516–28.

Thompson, William C. "Review of DNA Evidence in State of Texas v. Josiah Sutton (District Court of Harris County, Cause No. 800450)." (2003).

Vigilant, Linda, Mark Stoneking, Henry Harpending, Kristen Hawkes, and Allan C. Wilson. "African Populations and the Evolution of Human Mitochondrial DNA." *Science* 253, no. 5027 (1991): 1503–7.

Waldman, Shamam, Daniel Backenroth, Éadaoin Harney, Stefan Flohr, Nadia C. Neff, Gina M. Buckley, Hila Fridman, et al. "Genome-Wide Data from Medieval German Jews Show That the Ashkenazi Founder Event Pre-Dated the 14th Century." *Cell* 185, no. 25 (2022): 4703–16.

Wang, Ting, Lucinda Antonacci-Fulton, Kerstin Howe, Heather A. Lawson, Julian K. Lucas, Adam M. Phillippy, Alice B. Popejoy, et al. "The Human Pangenome Project: A Global Resource to Map Genomic Diversity." *Nature* 604, no. 7906 (2022): 437–46.

Weber-Lehmann, Jacqueline, Elmar Schilling, George Gradl, Daniel C. Richter, Jems Wiehler, and Burkhard Rolf. "Finding the Needle in the Haystack: Differentiating 'Identical' Twins Inpaternity Testing and Forensics by Ultra-Deep Next Generation Sequencing." *Journal of Forenic Sciences International Genetics* 9 (2014): 42–46.

Weir, B S. "Population Genetics in the Forensic DNA Debate." *Proceedings of the National Academy of Sciences* 89, no. 24 (1992): 11654–59.

"What Jennifer Saw." Episode. *Frontline* 15, no. 2. PBS, February 25, 1997.

White, T. J., T. Bruns, S. Lee, and J. Taylor. "Amplification and Direct Sequencing of Fungal Ribosomal RNA Genes for Phylogenetics." *PCR Protocols* (1990): 315–22.

Wisner, Mary, Erlich Henry, Shih Shelly, and Calloway Cassandra. "Resolution of Mitochondrial DNA Mixtures Using a Probe Capture Next Generation Sequencing System and Phylogenetic-Based Software." *Forensic Science International: Genetics* 53 (2021): 102531. Epub.

Index

For the benefit of digital users, indexed terms that span two pages (e.g., 52–53) may, on occasion, appear on only one of those pages.

Figures are indicated by *f* following the page number